# A SYSTEM OF USEFUL FORMULÆ,

ADAPTED TO THE PRACTICAL OPERATIONS OF

# LOCATING AND CONSTRUCTING RAILROADS:

A PAPER READ BEFORE THE BOSTON SOCIETY OF CIVIL ENGINEERS, DECEMBER, MDCCCXLIX.

---

By SIMEON BORDEN,
CIVIL ENGINEER.

---

BOSTON:
CHARLES C. LITTLE AND JAMES BROWN.
MDCCCLI.

BOSTON:
PRINTED BY WILLIAM CHADWICK,
EXCHANGE STREET.

# PREFACE.

The formulæ contained in the following pages were written, as their title indicates, for an original paper, which was read before and discussed by the Boston Society of Civil Engineers.

The Society, having accumulated a number of original papers, appointed a committee to examine and report upon the expediency of having them published. After much consideration, the committee reported that, with their limited resources and present necessities, it would be improper to incur the expense of printing, but recommended an early publication of this paper; and for that purpose the manuscript has been placed in the hands of the publisher.

Notwithstanding the obvious importance of constructing the curves of a railroad upon the best practical locations, and giving to their forms or alinement the greatest degree of regularity practicable, the investigation, or, which is more probable, the publication of anything like a system of convenient formulæ to aid the young engineer, and such others as have not had the advantage of a good mathematical education, in the proper performance of this character of work, it is believed has not yet found a place upon the shelves of our libraries or book-stores. To supply this deficiency in the library of the civil engineer, particularly the railroad engineer, is the object of the present paper.

It is not pretended that all the formulæ contained in this paper are original. The principles which have governed the investigations for computing the elements required for tracing curves, where their localities are such as to admit of the most simple and convenient methods, have, it is believed, been published, and are known by most engineers who have been engaged in the construction of railroads, since the commencement of the railroad system. Neither is it pretended that the system of formulæ is complete, or that it contains formulæ suited to every case that can arise. The writer can only say, that after considerable experience in the construction of railroads, he does not recollect a case presenting itself which would not be solved by some one of the formulæ; and it is believed that, with slight modifications, such as any geometer would be able, without difficulty, to make, they may be adapted to all common or ordinary cases.

Curves in a railroad, unless their radius be very large, are known to be objectionable; but the contour of the surface, the existence of valuable buildings, of streams, rivers, ponds, oceans, etc., in the line between the points which it is desired to connect, render the adoption of curves necessary. It is likewise a well-established fact, that the greater the degree of regularity and precision exercised in the construction of curves, the more safely and easily can trains be run over them.

The main objects of the formulæ are twofold; viz., that of enabling the engineer to mark out the curves of a railroad with the greatest degree of precision and convenience, and to locate them in situations the most desirable. To render this subject clear and perspicuous to every one who may have occasion to locate or mark out curves, upon railroads and other places, the paper is commenced with the investigation of the most simple problems, which are succeeded by the more intricate; each case being illustrated with diagrams, and accompanied by examples of computation.

The subject of switches and frogs being blended with the elements of turnout curves, has been considered in connection with them; and in their arrangement the same objects have been kept in view; and, for this end, each case has been likewise accompanied with a diagram and an example of computation.

To render the work more useful, there have been added formulæ for computing the cubic contents of excavations and embankments, and a formula for computing the difference in height to be observed in laying down the rails upon a railroad curve, based upon its radius and the velocity of the cars.

Boston, December, 1850.

# CONTENTS.

# ERRATA.

---

On page 77, ninth and tenth lines from top, for "$= 90° - Sw + \frac{1}{2}(180° - C)$" *read* "$(90° + Sw) - \frac{1}{2}(180° - C)$."

On page 87, seventh line from bottom, for "$= r + \frac{1}{2} h + d$" *read* "$r - \frac{1}{2} h + d$."

On page 113, second line from top, for "S 75° 08′ 35″·07 E" *read* "S 75° 08′ 35″·37 E."

On page 119, fourteenth line from bottom, for "We then discover" *read* "We then determine," etc.

# AN INVESTIGATION

OF

# USEFUL FORMULÆ,

ADAPTED TO

THE PRACTICAL OPERATIONS OF LOCATING AND LAYING OUT

RAILROAD CURVES.

[FIG. 1.]

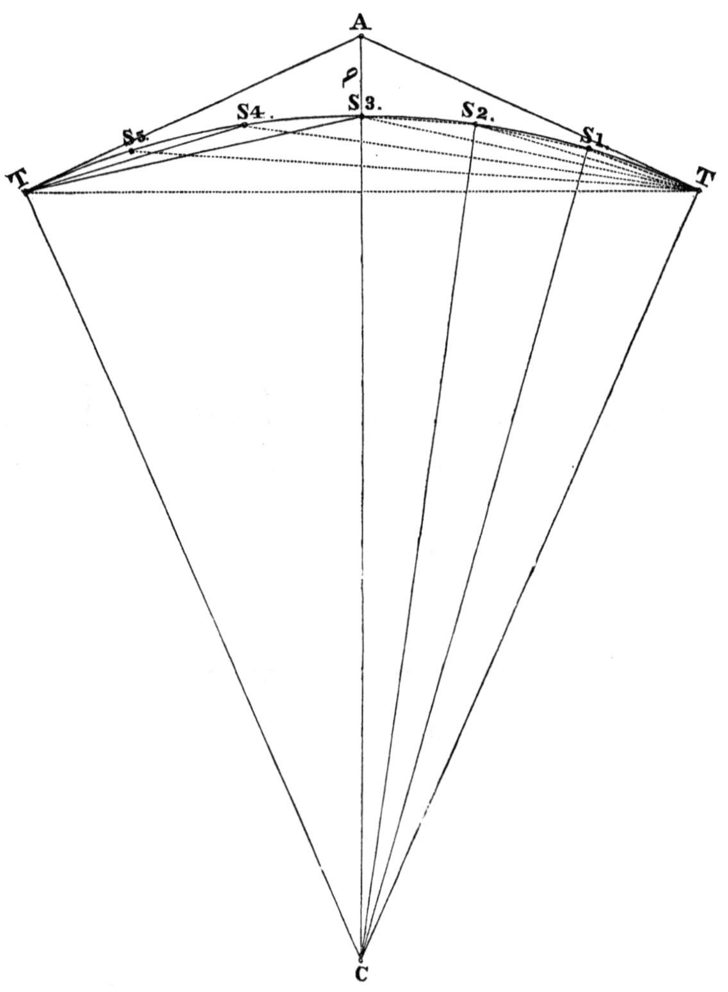

# An Investigation of Useful Formulæ,

ETC. ETC. ETC.

---

(1) COMMENCING with the most simple operations and forms of computations, we shall assume as the most simple of railroad curves, such as will unite two direct or straight lines of different bearings in such a manner that each of these lines will become tangents to said curve. Having run these straight or tangent lines to their intersections at A, (which intersection we shall hereafter term their apex, see Fig. 1,) and determined the magnitude of their angle, we then proceed to determine the most favorable location for the track. In order to do this, we examine the contour of the surface, and select the position we think the most favorable opposite the angle at A; then we run and measure from A to this most favorable place for the track before selected, in a direction that will bisect the angle A; that is, from A to $S_3$, and this distance we represent by $b$. We then proceed to ascertain a radius for the curve, which shall pass through the point $S_3$ and so run into the straight lines that they shall be tangents to said curve.

Putting $r$ for the radius in feet, or the unit of measure;

" C for the angle at the centre of the curve corresponding with the subtense of its arc $= (180° - A;)$

" $t$ for the distance from apex to the points of commencements of the curve, or where the direct line becomes a tangent to said curve, viz., from A to T, (See Fig.;)

We have $\sin. \frac{1}{4} C : b :: \sin. (\frac{1}{4} C + \frac{1}{2} A) : t = \frac{b \sin. (\frac{1}{4} C + \frac{1}{2} A)}{\sin. \frac{1}{4} C}$ and $\cos. \frac{1}{2} A : t :: \sin. \frac{1}{2} A : r = \frac{\tan. \frac{1}{2} A . b \sin. (\frac{1}{4} C + \frac{1}{2} A)}{\sin. \frac{1}{4} C} =$ $\tan. \frac{1}{2} A . b \cot. \frac{1}{4} C.$ (1)

Having thus determined the radius of the desired curve for uniting the aforesaid tangent lines, and the distance from their apex to their tangent points, or points of commencement of said curve, we will now proceed to investigate some of the most simple and practical methods of locating or laying out the same.

(**2**) We consider first what we shall call the method of deflections. To explain this operation, let us suppose the arc or curve to be divided into such equal parts as a chord of the length of the chain contemplated to be used will span. We have in the course of our practice generally used for this purpose a chain fifty feet in length; as, by using a short chain, the chords and the arcs (if the radius be of much magnitude) are nearly of the same length, which affords a great convenience in determining the deflecting angle corresponding to such short chords (consisting of fractions of the chain) as it will frequently be found desirable to use at the commencement and termination of curves, that we may be enabled to keep up a continuous notation of equi-distant stations.

Many engineers, I am aware, use a chain of 100 feet, which,

if the radius be not of considerable magnitude, will give a perceptible difference between the length of the chords and the corresponding arcs they span.*

To determine the angle of deflection corresponding to the length of the chain to be used, we represent the length of the chain by *ch;* the angle of deflection by D; and the angle at the centre of the curve, corresponding to its subtense, by C′.

Supposing T, $S_1$, = the chord *ch*, we have in the triangle C T $S_1$,

$$\tfrac{1}{2}\,(180^\circ - C') = \text{the angle } T = S_1\,;$$

Now as the angle A T C = a right angle, or 90°, we have

$$90^\circ - \left(\frac{180^\circ - C'}{2}\right) = D\,;$$

Multiplying by 2, $180^\circ - (180^\circ - C') = 2D$;

Subtracting 180°, and changing signs, we have

$$2D = C'\,;$$

Consequently, $$D = \tfrac{1}{2}\,C'\,; \qquad (2)$$

Bisecting *ch* we have $r : R :: \tfrac{1}{2}\,ch : \sin.\ \tfrac{1}{2}\,C' = \frac{\frac{1}{2}\,ch}{r} = \sin.\ D$ (3)

It sometimes happens that we wish to know the value of D at the commencement of our computations; expanding the foregoing,

$$\sin.\ D = \frac{\tfrac{1}{2}\,ch\ \cot.\ \tfrac{1}{2}\,A\ \tan.\ \tfrac{1}{2}\,C}{b} \qquad (4)$$

(**3**) Having thus determined the angle of deflection, we now are prepared for locating or marking out the curve.

Adjusting a good theodolite to the tangent point T, with its principal telescope (or the telescope by which angles are determined)

* In delineating a curve by the method of deflections, it will be inconvenient to make the stations further apart than the length of the chain used.

pointing in the direction of A, and with its watch telescope (all theodolites should be provided with watch telescopes) pointing to any convenient well-defined mark, lay off the angle D, and then stretching the chain from T, fix its terminus $S_1$ in range with the principal telescope, the point thus marked will be in the curve; then, laying off another angle of deflection which will read upon the instrument = 2D, stretch the chain from $S_1$ to $S_2$, fixing the terminus at $S_2$ in the range indicated by the main telescope; the points thus formed will also be in the curve. And in like manner we proceed to fix the points $S_3$, $S_4$, $S_5$, etc., until the curve is completed; or, as it more frequently happens, as long as the contour of the surface will permit us to see distinctly.

Let us now suppose an obstacle which will prevent seeing beyond $S_3$. We then remove our theodolite to that station; and, after having duly adjusted it, with its main telescope pointing at the station thus left, and the watch telescope to some convenient mark, lay off an angle equal to 180°, minus the sum of the deflections made at the first station, plus one deflection. Then, stretching the chain from the station where the theodolite is now adjusted, place the other terminus in range with the principal telescope, as heretofore described. Then, proceed in the manner above described, to lay off deflections and chords, until you connect with the straight or tangent lines, or as long as the contour of the country will admit, when the instrument must be again changed, and the like operations performed until the whole curve is completed. If the curve is not measured by whole chords = *ch*, the deflection for the fraction of *ch* will be to that of *ch* in the proportion the fraction bears to a whole; which, although not strictly correct, is sufficiently near for practice. We here remark that we do not recommend that more than 10 or

12 chords of 50 feet each should be laid down with the theodolite at one station, it being conducive to accuracy not to permit a great difference between the direction of the chord (*ch*) and the direction of the pointing of the telescope of the instrument.

(**4**) We have thus endeavored to describe a practical method of laying out one of the simplest of railroad curves which shall unite two straight lines having different bearings; but, as there is a great variety of methods to accomplish this end, which may be resorted to, some of which seem to possess peculiar adaptation to certain localities, I have thought it would not be uninteresting to describe some of them, believing a few hints of this kind would lead the new beginner to different modes of reasoning and investigation; and, if he possesses a tolerable knowledge of the elements of plain geometry, he will be able always to select, if not the method best adapted to the circumstances of the case, at least one well suited and convenient.

(**5**) If the curve be of large radius, (and curves cannot well be constructed with too large a radius,) and its location suits the contour or surface of the country, and the apex angle be large, a very convenient and accurate method of proceeding will be to divide the curve into a series of segments of some 500 or 600 feet each, as may be thought best; taking care as far as possible to make the terminus of each segment an even station;* it will, however, frequently happen that the number of the stations marking the tangent

* In the location and construction of a railroad, it has been found convenient in practice to divide the centre line into equal parts, technically called stations, which contain a given number of the units of measure used in the construction. In the United States, the foot has been taken for the unit of measure, and the railroad stations one hundred feet asunder.

points will contain a fraction, and of course in this case, the segments should contain, besides a number of whole chords, the fractional chord the case requires.

Having determined upon these preliminary considerations, we first calculate the relative position of points on the line between the tangent points and apex, which shall correspond to radii of the curve passing through the termini of the several segments; also the distance from the points upon the radii of the curve, together with the angles the radii made with the tangent lines; the points thus computed are readily determined and marked with a good degree of accuracy, and become instrument stations, from whence the intermediate stations are readily filled up as before described, by chords and deflections, with less liability to practical errors than the preceding method.

**(6)** If the curve be of large radius and the apex angle small, the distance of the tangent points from the apex and from the tangent lines to the curve, will become too great to be measured conveniently with a proper degree of accuracy; under this condition of the case, we divide the curve into a convenient number of segments, taking care as before to have their termini to correspond with the even stations.

Having thus determined upon the divisions, we compute the chords and angles corresponding therewith, and then proceed to lay off the angles with great care, and measure the chords with as great a degree of accuracy as is practicable, carefully marking the termini of each chord. If the chords and angles when laid down correspond with the tangent points and tangent lines, we proceed to put in the intermediate stations by adjusting the theodolite to the terminus of one of these chords; then, pointing to the other terminus, we proceed to lay down the intermediate chords and stations by the aid of deflections, in the manner described in the foregoing. It is usual to fill up the intermediate stations formed by two of the primitive chords without changing the instrument; having done this, we move our instrument to another terminus common to two chords, which have not been filled up with the intermediate stations, and in like manner we proceed until we complete the curve.

(7) If, in running our first chords until we have exhausted our computations, we do not find the work to correspond with a proper degree of accuracy to the tangent points, and tangent lines, we make a connection with the tangent points, and carefully ascertain the length of the chord, and the magnitude of the angle with the tangent line; and, with the elements thus obtained, we recompute the work, and determine the relative position of the tangent points, and fix them anew. We then proceed to perform the work of laying down the primitive chords a second time; when, if there have been no mistakes made, the work will prove practically accurate. It is very seldom that a second computation will be needed until the road is graded, when the measurement taken upon the graded surface will not be likely to be identical with the primitive measurement; then, a resurvey, and a re-establishment of the tangent points, become, if not absolutely necessary, desirable; but, if the first survey was performed with any tolerable degree of care, and the grading well finished, no difficulty will be experienced in laying the curve upon the graduated road-bed, without changing its radius.

**(8)** Having thus briefly described the principles which govern us in laying down simple curves, we would now introduce an example, accompanied by specimens of calculation for each particular case.

## PRACTICAL EXAMPLES OF CALCULATIONS SUITED TO THE CASES DESCRIBED IN THE FOREGOING PAGES.

Assuming $A = 160°$, then will $C = 180° - A = 20°$

$b = 132$ feet

$ch = 50$ feet

To find the deflection, we have (4) $\sin. D = \frac{\frac{1}{2} ch \cot. \frac{1}{2} A \tan. \frac{1}{4} C}{b}$

| | | | | |
|---|---|---|---|---|
| Thus, | $\frac{1}{2}$ A = 80° 00′ 00″ | …… | cot. = | 9·2463188 |
| | $\frac{1}{4}$ C = 5° 00′ 00″ | …… | tan. = | 8·9419518 |
| | $\frac{1}{2}$ $ch$ = 25 feet | …… | log. = | 1·3979400 |
| | $b$ = 132 feet co. ar. | …… | log. = | 7·8794261 |
| | D = 0° 10′ 02″.64 | …… | sin. = | 7·4656367 |

In this case it will be seen that D = 0° 10′ 02″.64 is an inconvenient angle to add or subtract, or even read upon the instrument; we therefore, to remedy this objection, adopt D = 0° 10′, which will not materially change the length of the radius or the location of the curve. This change requires that we base our calculations for determining the elements needed, upon D = 0° 10′; therefore, to find the radius we have

$$\text{Sin. } D : \tfrac{1}{2}\, ch :: R : r = \frac{\frac{1}{2} ch}{\sin. D} \qquad (5)$$

and to find $t$ we have

$$\text{Cos. } \tfrac{1}{2}\, C : r :: \sin. \tfrac{1}{2}\, C : t = \tan. \tfrac{1}{2}\, C \,.\, r \qquad (6)$$

(representing by $t$ the distance from the apex to the tangent point, or from A to T on the diagram.)

| | | | |
|---|---|---|---|
| Thus, ........ | $\frac{1}{2}\ ch = 25$ feet | .... log. = | 1·3979400 |
| | D = 0° 10′ co. ar. | .... sin. = | 2·5362745 |
| Consequently, | $r$ = 8594·38 feet | ........ log. = | 3·9342145 |
| Then, ........ | $\frac{1}{2}$ C = 10° 00′ 00″ | .... tan. = | 9·2463188 |
| | $t$ = 1515·42 feet | ........ log. = | 3·1805333 |

Having thus ascertained the radius $r = 8594 \cdot 38$ feet, and the distance from apex to tangent point, ($t = 1515 \cdot 42$ feet,) we now proceed to find $b$. By analogy we have

$$\text{Sin.}\left(\tfrac{1}{4}\,C + \tfrac{1}{2}\,A\right) : t :: \sin. \tfrac{1}{4}\,C : b = \frac{t \sin. \frac{1}{4}\,C}{\sin.\left(\frac{1}{4}\,C + \frac{1}{2}\,A\right)} = t \tan. \tfrac{1}{4}\,C \quad (7)$$

| | | | |
|---|---|---|---|
| Then, ...... | $t$ = as above, | .... log. = | 3·1805333 |
| | $\frac{1}{4}$ C = 5° 00′ 00″ | .... tan. = | 8·9419518 |
| | $b$ = 132·58 feet | ... log. = | 2·1224851 |

At commencement we assumed } $b$ = 132·00 feet

Difference, .... = 0·58 feet

Thus it appears that the change of the radius to cause it to correspond with D = 0° 10′ will only change the location of the curve from the position designed for it 0·58 feet; a quantity too small to be generally accounted anything in choosing the position of a curve.

We now proceed to find the length of the curve. Putting $r''$ for an arc in seconds = radius, and C″ for the number of seconds contained in the centre angle which measures the curve, and $a$ the arc subtending the angle C″, we have by analogy

$$r'' : r :: C'' : a = \frac{r\,C''}{r''} \quad (8)$$

| | | | | |
|---|---|---|---|---|
| Thus, | $r$ = 8594.38 feet | | log. = | 3.9342145 |
| | C″ = 7200″ | | log. = | 4.8573325 |
| | $r''$ = | co. ar. | log. = | 4.6855749 |
| | $a$ = 3000.00 feet | | log. = | 3.4771219 |

[Fig. 2.]

A
Station 155
2
arc=7.50 Stations
Ch.749.76 feet
Station 162.50
3
Station 147.50
1
Station 170.38
T'
arc=7.88 Stations
Ch.787.72 feet
Station 140.38
T
arc=7.12 Stations
Ch.711.79 feet
No.1
No.2
No.3
No.4
C

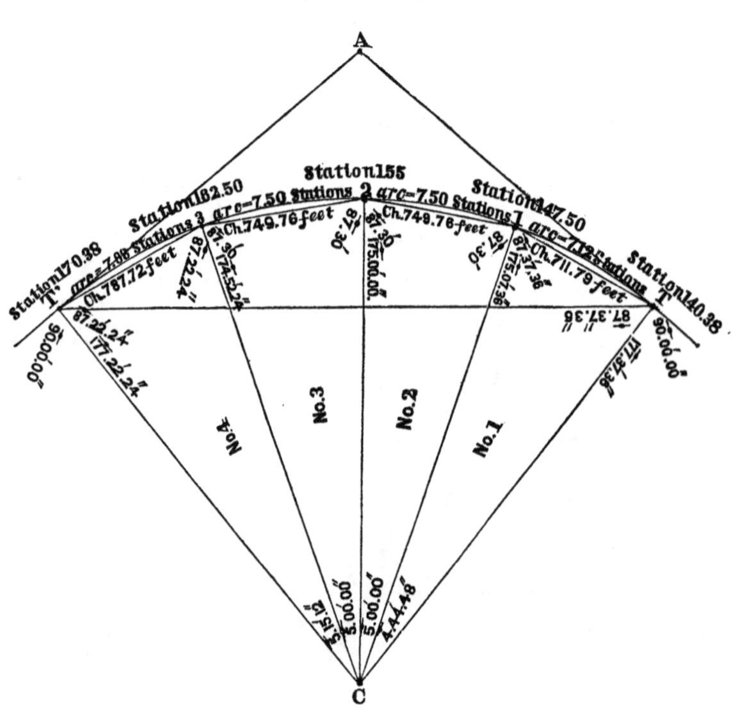

Now, that we may show this matter as complicated as it is generally found in practice, let us suppose the pin at T marking the first tangent point, to be numbered 140·38; the integer representing the whole number of the stations, and the decimals the fractional part of the space beyond station 140. We have found the length of the arc $a$ equal 30 whole stations; which, added to 140.38, gives 170·38 for the number of the other tangent station T'.

(**9**) Let us now proceed to show the method of laying out the curve by the method of a series of long chords corresponding to about 750 feet of the arc, which we afterwards fill up by simple deflections and 50 feet chords $= ch$.

In the first place we have supposed the first tangent pin to bear the number 140·38; if we now add 7·12 stations, it will make the number 147.50; therefore our first chord will extend from station 140·38 to station 147·50. We now add 7·50 stations to station 147·50, which makes the number 155; therefore our second chord extends from station 147·50 to station 155. We then add 7·50 stations to station 155, which increases the number to 162·50; therefore our third chord extends from station 155 to 162·50. We now add 7·88 stations to 162·50, which increases the number to 170·38, and brings us to the tangent point T'.

Having assumed $ch = 50$ feet, (which is practically the same length of the arc it spans when the radius is of considerable length,) we must now proceed to determine the length of the several long chords we have divided the arc into. (See the diagram on the opposite page.)

Our first long chord corresponds to 712 feet of arc. We now propose to find the centre angle which it subtends by the following formula.

Representing the arc by $a$ and the centre angle in seconds by C″, we have

$$r : r'' :: a : C'' = \frac{r'' a}{r} \qquad (9)$$

and further we have

$$\text{Cos. } \tfrac{1}{2}\, C'' : r :: \text{sin. } C'' : Ch^{*} = \frac{\text{sin. } C'' r}{\text{cos. } \frac{1}{2} C''} \qquad (10)$$

| | | | | |
|---|---|---|---|---|
| Thus, | $r$ = 8594·38 feet | co. ar. | log. = | 6·0657855 |
| | $a$ = 712·00 feet | | log. = | 2·8524800 |
| | $r''$ = | | log. = | 5·3144251 |
| | C″ = 17088″ | | log. = | 4·2326906 |

| | | | |
|---|---|---|---|
| C″ reduced to degrees and minutes = 4° 44′ 48″ | | sin. = | 8·9177692 |
| ½ C″ " " " = 2° 22′ 24″ | ar. co. | cos. = | 0·0003719 |
| $r$ ........................................ | | log. = | 3·9342145 |
| First chord = $Ch$ = 711·79 feet | = | log. = | 2·8523556 |

The arc corresponding to the succeeding chord is 750 feet long, and is composed of 15 whole deflections, and each deflection being 10′, the centre angle will be equal to twice that number of deflections, or 300′, which amounts to 5° 00′ 00″.

Now, by formula (10) we have

* We distinguish the primitive or long chords by $Ch$ in contradistinction with the 50 feet or deflecting chords, which are represented by $ch$.

| | | |
|---|---|---|
| C″ in degrees = 5° 00′ 00″ | sin. = | 8·9402960 |
| ½ C″ in degrees = 2° 30′ 00″ co. ar. | cos. = | 0·0004135 |
| r = ............................ | log. = | 3·9342145 |
| Second chord = 749·763 feet | log. = | 2·8749240 |

Third chord, of course, will be of the same length, viz., 749·763.

The arc corresponding to the fourth chord is 788 feet long, and we find C″ and *Ch* by formula (9) and (10)

| | | | |
|---|---|---|---|
| Thus, .... r ............ | co. ar. | log. = | 6·0657855 |
| a = 788 feet | | log. = | 2·8965262 |
| r″ ............ | | log. = | 5·3144251 |
| C″ = 18912″ | | log. = | 4·2767368 |

| | | |
|---|---|---|
| C″ reduced to degrees and minutes = 5° 15′ 12″ | sin. = | 8·9617037 |
| ½ C″ " " " " " = 2° 37′ 36″ co. ar. | cos. = | 0·0004554 |
| r ......................... = | log. = | 3·9342145 |
| Fourth chord = *Ch* = 787·723 feet | log. = | 2·8963736 |

We have not attempted to make the foregoing calculations strictly exact; our angles being always taken to correspond with the nearest second, which in most cases gives a greater degree of accuracy than we can practically execute.

**(10)** Having thus computed the centre angles and chords corresponding with the proposed division of the arc, we will endeavor to give the method of laying out the work.

In the first place we will determine the angles at the tangent points, and the several stations which are to mark the termini of the divisions of chords.

Commencing at the tangent point numbered 140·38, our first operation is to determine the angles in the several triangles, viz.; T C 1, 1 C 2, 2 C 3, 3 C T'. (See Fig. 2.)

The angle C'', which we have determined, will correspond to C, and in triangle T C 1, C = 4° 44′ 48″, (see section 9,) each of these triangles being isosceles, and having computed their centre angles, we have only to deduct them from 180°, and half the remainder will be the value of each of the remaining angles. Now, calling triangle T C 1 No. 1, and 1 C 2 No. 2, and 2 C 3 No. 3, and 3 C T No. 4, we proceed to find their angles in the order we have named them.

DETERMINATION OF THE ANGLE AT FIRST TANGENT POINT, OR T.

The angle at centre of curve for triangle No. 1 = 4° 44′ 48″ and $\frac{180° - 4° 44' 48''}{2}$ = 87° 37′ 36″ = angle at T, or at station 140·38 and 147·50.

" angle at centre of curve for triangle No. 2 = 5° 00′ 00″ and $\frac{180° - 5° 00' 00''}{2}$ = 87° 30′ = angle at stations 147·50 and 155.

" angle at centre of curve for triangle No. 3 = 5° 00′ 00″ and $\frac{180° - 5° 00' 00''}{2}$ = 87° 30′ = angle at stations 155 and 162·50.

" angle at centre of curve for triangle No. 4 = 5° 15′ 12″ and $\frac{180° - 5° 15' 12''}{2}$ = 87° 22′ 24″ = angle at stations 162·50 and 170·38.

Having thus prepared the angles for the several stations above named, and for the purpose of rendering our description easier to be understood, we arrange them in their order, as in the foregoing diagram.

**(11)** Having thus represented our work as shown in the diagram, we now proceed to ascertain the angles at the termini of the several chords. Thus,

| | | | |
|---|---|---|---|
| At T, or station, .... | 140·38 | = 90° + 87° 37′ 36″ | = 177° 37′ 36″ |
| " 1, " .... | 147·50 | = 87° 37′ 36″ + 87° 30′ | = 175° 07′ 36″ |
| " 2, " .... | 155·00 | = 87° 30′ + 87° 30′ | = 175° 00′ 00″ |
| " 3, " .... | 162·50 | = 87° 30′ + 87° 22′ 24″ | = 174° 52′ 24″ |
| " T′, " .... | 170·38 | = 87° 22′ 24″ + 90 | = 177° 22′ 24″ |

Having ascertained the angles for each station, which, for convenience, we write upon the several radii connecting said stations with the centre of the curve in the diagram, we proceed to compute the length of the several chords which span their respective arcs; commencing with triangle T C 1, which is called No. 1, and pursuing the calculations in the order shown in the diagram. Thus,

TRIANGLE No. 1.

| | | | |
|---|---|---|---|
| T | = 87° 37′ 36″ .... co. ar. .... | sin. | = 0·0003727 |
| $r$ | = 8594·38 feet | log. | = 3·9342145 |
| C | = 4° 44′ 48″ | sin. | = 8·9177692 |
| | *Ch* = 711·79 feet .... | = log. | = 2·8523564 |

TRIANGLES Nos. 2 and 3.

| | | | |
|---|---|---|---|
| 1 | = 87° 30′ 00″ .... co. ar. .... | sin. | = 0·0004135 |
| $r$ | = | log. | = 3·9342145 |
| C | = 5° 00′ 00″ | sin. | = 8·9402960 |
| | *Ch* = 749·76 feet .... | = log. | = 2·8749240 |

TRIANGLE No. 4.

| | | | |
|---|---|---|---|
| 3 | = 87° 22′ 24″ .... co. ar. .... | sin. | = 0·0004565 |
| $r$ | = | log. | = 3·9342145 |
| C | = 5° 15′ 12″ | sin. | = 8·9617037 |
| | *Ch* = 787·72 feet .... | = log. | = 2·8963747 |

**(12)** Having thus prepared our work, we proceed to adjust the theodolite to station T, or according to the locating stations, to No. 140·38, with its principal telescope pointing in the direction of the line of the road; which, it is presumed, has been properly marked. Then, laying off an angle of 177° 37′ 36″, we measure in the direction indicated by the telescope, 711·79 feet to 1 or to station 147·50; then, moving the instrument to 147·50, and pointing the principal telescope to T, we lay off an angle of 175° 07′ 36″, and measure in the direction indicated 749·76 feet to 2 or station 155. Then, moving the instrument to station 2 and pointing at 1, we lay off an angle of 175°, and measure in the direction indicated 749·76 feet to 3 or to station 162·50. Then, moving the instrument to 3 and pointing at 2, we lay off an angle of 174° 52′ 24″, and measure in the direction indicated 787·72 feet to T′ or to station 170·38; which, if our angles and measures bring us direct to T′ or near to it, we presume the work to be correctly done. We should, however, before pronouncing the work correct, place our instrument at T′, and pointing the telescope to 3, lay off the angle with the line of the road, and if this agrees with the computed angle, I think we may then, without hesitation, pronounce the work correct.

But, if our angles and measures do not bring us direct to T′ or near by it, we then point our telescope to T′, and ascertain the angle indicated by the instrument, and measure the distances as correctly as we can, which we duly note down in our field book. We then move to T′ with our instrument, and pointing its telescope to 3, we measure the angle with the line of the road, which we also note in our field book.

With the data thus obtained, we proceed to recompute the

elements of a curve that will unite the two lines without materially varying the location of the track from the points which we have just fixed.

**(13)** In order to show practically the performance of these operations, we will make up the following as the field notes of a survey for locating the curve above described.

Commencing at T with instrument pointing in the direction of the road,

| | | | | | |
|---|---|---|---|---|---|
| | we laid off the angle | 177° 37′ 36″ | and measured 711·79 | feet to station | 1 |
| At station 1 | we laid off the angle | 175° 07′ 36″ | and measured 749·76 | feet to station | 2 |
| " 2 | " " | 175° 00′ 00″ | " 749·76 | " | 3 |
| " 3 | " " | 175° 00′ 00″ | " 751·51 | " | T |
| " T | " " | 177° 14′ 48″ | in the direction of road. | | |
| | Sum, . . . . | 880° 00′ 00″ | | | |

Having obtained the field notes of our traverse, our first operation will be to deduce from them the angle at apex, and at the centre of the curve. We here remark that the sum of the angles at apex and at the centre of the curve, always amount to 180°, and of course one must be a supplement to the other.

To ascertain either of the angles, viz., at the apex, or at the centre, a variety of formulæ might be deduced, but it is presumed the following is as convenient as any; viz., subtract the sum of all the angles from as many times 180° as there are angles, and the remainder will be the angle at the centre of the curve.

It will be seen that we have noted in our field book five angles, whose sum amounts to 880°; now $5 \times 180 = 900$; and $900 - 880$

$= 20° =$ the angle at the centre of the curve, which compares with the angle we had formerly ascertained.

Our next step is to ascertain the relative position of the present points T and T′ with respect to A, and also the position they should occupy to suit the radius we have heretofore deduced.

We know of no more convenient method of determining this problem, than by working up the traverse, (as the ship captains call it,) and for that purpose we will assume the line of the road extended from T to apex, whatever may be its direction, as bearing due north, and predicate the bearings of the other lines upon it, as indicated by the angles. Thus,

Angle at T = 177° 37′ 36″ and 180° — 177° 37′ 36″ leaves 2° 22′ 24″ N. W. to 1
" " 1 = 175° 07′ 36″ = 352° 45′ 12″ and (2 × 180°) — 352° 45′ 12″ = 7° 14′ 48″ " 2
" " 2 = 175° 00′ 00″ = 527° 45′ 12″ " (3 × 180°) — 527° 45′ 12″ = 12° 14′ 48″ " 3
" " 3 = 175° 00′ 00″ = 702° 45′ 12″ " (4 × 180°) — 702° 45′ 12″ = 17° 14′ 48″ " T′
" " T′ = 177° 14′ 48″ = 880° 00′ 00″ " (5 × 180°) — 880° 00′ 00″ = 20° 00′ 00″ "

being the direction of the road.

Computing the northings and westings of the foregoing traverse, we have

| | | | | | |
|---|---|---|---|---|---|
| No. 1. N. W. 2° 22′ 24″ | sin. = 8·6171119 | | cos. = 9·9996273 |
| 711·79 feet | log. = 2·8523458 | | log. = 2·8523458 |
| 29·475 " | log. = 1·4694577 | 711·17 feet | log. = 2·8519731 |
| No. 2. N. W. 7° 14′ 48″ | sin. = 9·1008572 | | cos. = 9·9965171 |
| 749·76 feet | log. = 2·8749223 | | log. = 2·8749223 |
| 94·576 " | log. = 1·9757795 | 743·77 feet | log. = 2·8714394 |

| | | | | |
|---|---|---|---|---|
| No. 3. N. W. 12° 14′ 48″ | sin. = 9·3265833 | | cos. = 9·9900028 |
| 749·76 feet | log. = 2·8749223 | | log. = 2·8749223 |
| 159·04 " | log. = 2·2015056 | 732·70 feet | log. = 2·8649251 |
| No. 4. N. W. 17° 14′ 48″ | sin. = 9·4720042 | | cos. = 9·9800203 |
| 751·51 feet | log. = 2·8759348 | | log. = 2·8759348 |
| 222·814 " | log. = 2·3479390 | 717·72 feet | log. = 2·8559551 |

Then, summing up the computed northings and westings, we have as follows:

| | Bearing. | Distance. | Northing. | Westing. |
|---|---|---|---|---|
| | ° ′ ″ | | | |
| From T to No. 1 = | N. W. 2 22 24 | 711·79 | 711·17 | 29·475 |
| " 1 " 2 = | " 7 14 48 | 749·76 | 743·77 | 94·576 |
| " 2 " 3 = | " 12 14 48 | 749·76 | 732·70 | 159·040 |
| " 3 " 4 = | " 17 14 48 | 751·51 | 717·72 | 222·814 |
| | Total, = | .......... | 2905·36 | 505·905 |

Having summed up the traverse, we now proceed to find the bearing and distance from T to T′.

Putting N = the sum of the northings, and calling it the cos.;
W = " " westings, " " sin.;
C = the bearing sought;
D = the distance from T to T′.

We then have by analogy the following formulæ:

$$N : W :: R : \tan. C = \frac{W}{N} \qquad (A)$$

$$\text{and } \sin. C : W :: R : D = \frac{W}{\sin. C} \qquad (B)$$

$$\text{or } \cos. C : N :: R : D = \frac{N}{\cos. C} \qquad (C)$$

[FIG. 3.]

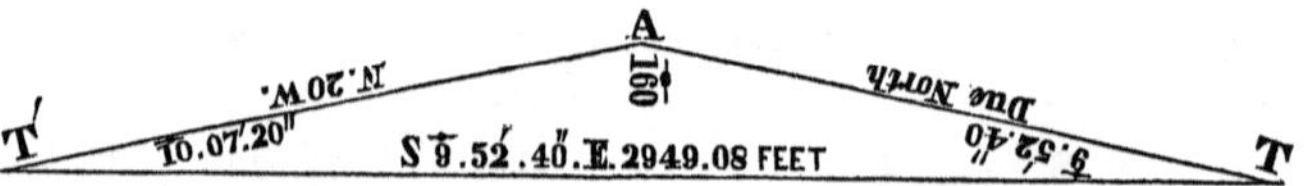

Performing the computations indicated, we have

| | | | |
|---|---|---|---|
| W = 505·905 | | ........ log. = | 2·7040690 |
| N = 2905·36 | co. ar. | ........ log. = | 6·5368001 |
| C = 9° 52′ 40″ | | ........ tan. = | 9·2408691 |
| C = 9° 52′ 40″ | co. ar. | ........ sin. = | 0·7656168 |
| W = | | ........ log. = | 2·7040690 |
| D = 2949·08 feet | | ........ log. = | 3·4696858 |

We now have in the triangle A T T′, the side T T′ and the data for finding the unknown angles.* Then, to find the distances A T and A T′ we have

$$\text{sin. A} : \text{D} :: \text{sin. T} : \text{A T}' = \frac{\text{D sin. T}}{\text{sin. A}}$$

$$\text{and sin. A} : \text{D} :: \text{sin. T}' : \text{A T} = \frac{\text{D sin. T}'}{\text{sin. A}}$$

To prevent confusion in our diagram, or to render our work more plain, we reconstruct the figure of the triangle A T T′. (See figure on preceding page.)

To perform the computations indicated, we have

| | | | |
|---|---|---|---|
| A = 160° 00′ 00″ | co. ar. | ........ sin. = | 0·4659483 |
| D = 2949·08 feet | | ........ log. = | 3·4696861 |
| T′ = 10° 07′ 20″ | | ........ sin. = | 9·2448918 |
| A to T = 1515·396 feet | | ........ log. = | 3·1805262 |
| Sum of logs. of A and D = | | ........ log. = | 3·9356344 |
| T = 9° 52′ 40″ | | ........ sin. = | 9·2343832 |
| A to T′ = 1479·7 feet = | | ........ log. = | 3·1700176 |

* The data for finding the unknown angles are the relative bearings of the sides of the triangles, as found in the preceding computations.

Having thus computed the distance from apex to the points T and T′, we will now ascertain the length those lines should be to suit the contemplated radius; thus we have

$$\text{Cos. } \tfrac{1}{2} \text{ C} : r :: \text{sin. } \tfrac{1}{2} \text{ C} : t = \text{tan. } \tfrac{1}{2} \text{ C}.r$$

| | | | |
|---|---|---|---|
| $r$ | = 8594·38 feet | ........ log. | = 3·9342145 |
| $\frac{1}{2}$ C | = 10° 00′ 00″ | ........ tan. | = 9·2463188 |
| $t$ | = 1515·42 feet | ........ log. | = 3·1805333 |

$t$ in the present case, as in our former notation, represents the distance required from A to T and also from A to T′, to suit the contemplated curve, which in the present instance, has a radius of 8594·38 feet, and an apex angle A = 160°.

It appears from the above computations that T should be moved from the apex 1515·42 — 1515·396 = ·024 feet, which amount in practice is so small that we should consider T correctly located, and doubtless our calculations would have given its location exact, had we been careful in the management of the fractions; but it is not so with T′. The computations show that T′ should be moved from apex 1515·42 — 1479·17 = 36·25 feet. After having moved T′ from apex 36·25 feet, in the direction of the line of the road, there will be no doubt, if the previous traverse upon which the calculations have been based, has been correctly measured, that the contemplated curve could be accurately located between the tangent points as corrected.

The above calculations have been based upon a traverse consisting of chords of an arc correctly laid down, with the exception of the last course; but, had the traverse been a random one, the results arrived at would have been equally exact, with only this difference, that the points T and T′ would not probably have been found so near their proper location.

The practice explained above will be found useful in locating curves by the side of rivers, ponds, oceans, mountains, rough surfaces, etc.; in short, wherever it is found inconvenient to run the direct lines to their intersections, or to apex, and to measure therefrom to the points T and T′. In running traverses for the purpose of obtaining the elements for the location of curves under the conditions suggested above, it will be found convenient, if not absolutely necessary, that some portion of the traverse should be so made as to give the relative position of such points as the contour of the surface or other considerations may render it desirable that the location should pass through. We now proceed to give a practical example.

**(14)** The following practical example supposes the direct or straight lines to be united by the curve, to have been located and marked by some convenient device, and the angles given below to have been measured by a common theodolite, and the lines by a chain, thus:*

Field notes of a traverse for obtaining the elements of a curve for uniting the lines T and T′.

Commencing at station 0, corresponding in the diagram to T,

* This remark would seem to be uncalled for, as it can make no difference in the computation how the lines and angles are obtained; but it frequently happens that the traverse is obtained by means of a triangulation, which would sometimes present the matter in a different form. The writer has had a number of cases that could not have been well performed in any other manner.

with telescope pointing in the direction of the located line, we measured as follows, viz.:

| | ANGLES. | | DISTANCES. | |
|---|---|---|---|---|
| | ° ′ ″ | | | |
| Station 0 = | 169 29 45 | = | 1200 feet to station 1 | The curve should pass through or near this point, which of course governs the radius. |
| " 1 = | 170 00 15 | = | 900 " " 2 | |
| " 2 = | 175 04 30 | = | 750 " " 3 | |
| " 3 = | 164 40 10 | = | 1525 " " 4 | corresponding in diagram to T′ |
| " 4 = | 140 45 20 | = | in the direction of the located line of the road. | |
| | 820 00 00 | | | |
| 180 × 5 = | 900 00 00 | | | |
| | 80 | = | angle at centre of curve. | |

Assuming the line from T to A as bearing due north, whatever its course may be, we deduce the following relative courses for the several lines of the traverse, and by formula (A) ascertain the relative bearing from T to T′.

| STATION | BEARINGS. | DISTANCES. | NORTHINGS. | WESTINGS. | |
|---|---|---|---|---|---|
| 0 to 1 = | N. W. 10° 30′ 15″ | 1200 | 1179·890 | 218·7685 | |
| 1 " 2 = | " 20° 30′ 00″ | 900 | 843·005 | 315·1866 | |
| 2 " 3 = | " 25° 25′ 30″ | 750 | 677·361 | 321·9970 | |
| 3 " 4 = | " 40° 45′ 20″ | 1525 | 1163·197 | 995·5700 | |
| | | | 3863·453 | 1851·5221 | log. = 3·2675289 |
| | | | | 3863·453 | log. = 3·5869756 |
| Relative bearing from T to T′ = N. W. 25° 36′ 20″·22 | | | | | tan. = 9·6805533 |

Having thus obtained the bearing from T to T′, we now proceed to compute the distance; by formula (B) and (C) we have

$$\text{Sin. C} : \text{W} :: \text{R} : \text{D} = \frac{\text{W}}{\text{sin. C}} \quad \text{or}$$

$$\text{Cos. C} : \text{N} :: \text{R} : \text{D} = \frac{\text{N}}{\text{cos. C}}$$

In these analogies, C represents the course from T to T′; W the westing; N the northing; D the distance.

We frequently, as we shall in the present instance, use both formulæ, for the purpose of proof.

| | | | |
|---|---|---|---|
| W = 1851·5221 | log. = 3·2675289 | N = 3863·453 | log. = 3·5869756 |
| C = 25° 36′ 20″.22 | sin. = 9·6356588 | C = 25° 36′ 20″.22 | cos. = 9·9551055 |
| D = 4284·2 feet | log. = 3·6318701 | D = 4284·2 feet | log. = 3·6318701 |

Our next step in practice is to ascertain the distances from T and T′ respectively to the point where the direct lines would intersect; or, in other words, to the apex.

We have ascertained the angle of the centre of the curve to be 80°. Of course the angle at apex will be 100°. The bearings which we have ascertained also indicate the angles; thus, in the imaginary triangles we are about to solve, we have supposed the line from T to A to bear due north. Then, by computation, the line from T to T′ bears N. W. 25° 36′ 20″.22, which gives the angle at T the same number of degrees as the bearing. From the traverse or the table of angles, in our field notes, we deduce the bearing of the located line from T to be N. W. 80° 00′ 00″.

| | | |
|---|---|---|
| These bearings indicate the following angles, viz., | at A = | 100° 00′ 00″.00 |
| As before stated, at | T = | 25° 36′ 20″.22 |
| | T′ = | 54° 23′ 39″.78 |
| Proof, | Sum = | 180° 00′ 00″.00 |

With these angles, and the distance from T to T′ = D, the distances T A and T′ A are readily found; thus,

[Fig. 4.]

A
100
T′
N. 80. W.
54.23.39.78
3
2
1
Due North
N 25.36.20.22 W.
25.36.20.22
T″
T T″
C

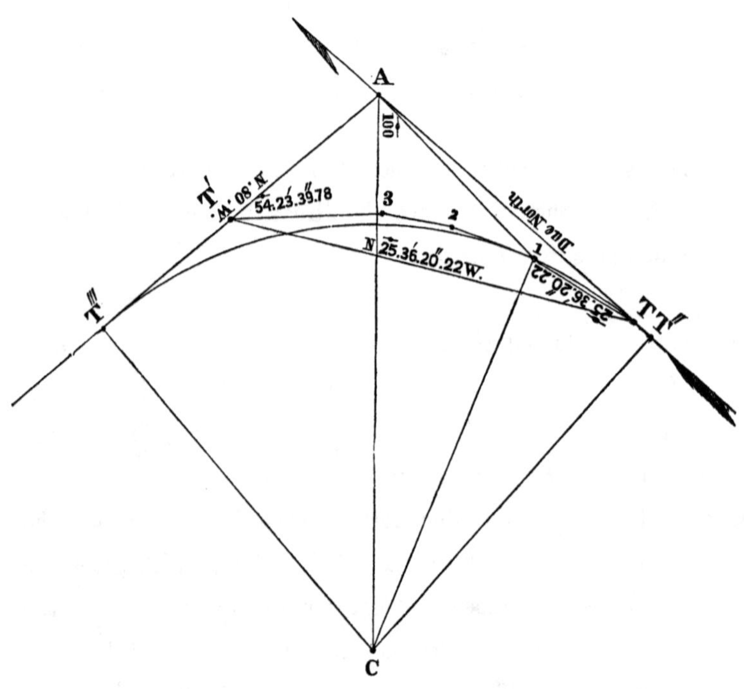

$$\left.\begin{array}{r} \text{sin.} A : D :: \text{sin.} T' : A T \\ \text{and sin.} A : D :: \text{sin.} T : A T' \end{array}\right\} \qquad (10)$$

| | | | | |
|---|---|---|---|---|
| A | = 100° 00′ 00″ | ar. co. | ........ sin. = | 0·0066485 |
| D | = 4284·2 feet | | ........ log. = | 3·6318701 |
| T | = 54° 23′ 39″.78 | | ........ sin. = | 9·9101139 |
| A T | = 3536·98 feet | | ........ log. = | 3·5486325 |

| | | | | | |
|---|---|---|---|---|---|
| Again, | A | = 100° 00′ 00″ | co. ar. | ........ sin. = | 0.0066485 |
| | D | = 4284·2 feet | | ........ log. = | 3·6318701 |
| | T | = 25° 36′ 20″.22 | | ........ sin. = | 9·6356588 |
| | A T′ | = 1880·085 feet | | ........ log. = | 3·2741774 |

For the triangle A $T_1$ (see diagram) we have by our measurement and computations the sides A T = 3536·98 feet, and 1 T = 1200 feet with their included angle = 10° 30′ 15″ to find A 1 and the unknown angles.

For convenience in the enunciation of the formula, let A T = $a$, 1 T = $b$, C = the given angle, and A and B the unknown angles; A representing the unknown angle opposite the side $a$, and B the unknown angle opposite the side $b$. We then have $a + b$ : $a \backsim b$ :: tan. ½ (180° — C) : tan. ½ (A ∽ B) and ½ (180° — C) + ½ (A ∽ B) = the angle opposite the longest side, and ½ (180°— C) — ½ (A ∽ B) = the angle opposite the shortest side.

In the calculations which follow we change the symbols from those given in the formula, so as to have them conform to the letters and figures given upon the diagram.

| | | | | |
|---|---|---|---|---|
| A T | = 3536·98 | Given angle T = | | 180° 00′ 00″<br>10° 30′ 15″ |
| 1 T | = 1200 | | | 2) 169° 29′ 45″<br>84° 44′ 52″.5 |
| Sum | = 4736·98 | co. ar. ........ | log. | = 6·3244984 |
| Diff. | = 2336·98 | ........ | log. | = 3·3686550 |
| ½ (180°— T) | = 84° 44′ 52″·50 | ........ | tan. | = 1·0365722 |
| ½ (A ∽ 1) | = 79° 26′ 43″·14 | | tan. | = 0·7297256 |
| ∠ at 1 | = 164° 11′ 35″·64 | | | |
| ∠ at A | = 5° 18′ 09″·36 | | | |
| Given ∠ at T | = 10° 30′ 15″·00 | | | |
| | 180° 00′ 00″·00 | | | |

| | | | | |
|---|---|---|---|---|
| ∠ at 1 | = 164° 11′ 35″·64 | co. ar. .... | sin. | = 0·5648026 |
| A T | = 3536·98 feet | .... | log. | = 3·5486326 |
| T | = 10° 30′ 15″ | .... | sin. | = 9·2608034 |
| A 1 | = 2367·21982 feet | .... | log. | = 3·3742386 |

| | | | | |
|---|---|---|---|---|
| A | = 5° 18′ 09″·36 | co. ar. .... | sin. | = 1·0342539 |
| T 1 | = 1200 feet | .... | log. | = 3·0791812 |
| T | = 10° 30′ 15″ | .... | sin. | = 9·2608034 |
| Proof, | = 2367·21982 feet | .... | log. | = 3·3742385 |

Having obtained all the elements of the triangle A 1 T we represent the side A 1 as found above in feet, or in general in the unit of measure by $b_m$, and as we find it in proportion to the radius of the curve = unity, by $b$.

As it becomes convenient generally to use the letters standing against the angles in each triangle, and as some of them are common at least to three triangles, it becomes necessary to occasionally accent some of them, that we may understand their different values in the investigation. We have throughout our investigations represented the apex angle by A. We shall continue to give to A that value in the following investigation, with the exception of angle A in the triangle A 1 T (which in general will not be a multiple of the apex angle) we shall therefore represent it by A′.

Commencing our investigation with the triangle A T″ C, and representing T″ C the radius of the curve, by unity, and the radius of the tables by R, we have

$$\text{Sin. } \tfrac{1}{2} A : 1 :: R : A C = \frac{1}{\sin. \frac{1}{2} A} \tag{11}$$

In the triangle A C 1 we have C 1 = the radius of the curve = 1; therefore, representing the angle at 1 by G, we have

$$1 \,.\, \sin. (\tfrac{1}{2} A - A') :: A C : \sin. G = \frac{\sin. (\frac{1}{2} A - A')}{\sin. \frac{1}{2} A} \tag{12}$$

For convenience representing the line A C by $d$, we have

$$\text{Sin. } G : d :: \sin. [(\tfrac{1}{2} A - A') + G] : b = \frac{d \sin. [(\frac{1}{2} A - A') + G]}{\sin. G} \tag{13}$$

then $$b : bm :: 1 : r = \frac{bm}{b} \tag{14}$$

where $r$ represents the radius of the curve in the unit of measure.

Or, probably the following formula would be rather more simple for calculation than the above, (13) viz.:

$$\text{Sin. } (\tfrac{1}{2} A - A') : 1 :: \sin. [(\tfrac{1}{2} A - A') + G] : b = \frac{\sin. [(\frac{1}{2} A - A') + G]}{\sin. \frac{1}{2} A - A'} \tag{13'}$$

then, as above $$b : bm :: 1 : r = \frac{bm}{b} \tag{14'}$$

where $r$ represents the radius of the curve in the unit of measure.

Example of calculation, formula (12) $\sin. G = \frac{\sin. (\frac{1}{2} A - A')}{\sin. \frac{1}{2} A}$

We found A by the reduction of the foregoing traverse (see preceding pages) = 100°, also A′ (noted as A) in the triangle A 1 T = 5° 18′ 09″.36. Therefore,

| | | | |
|---|---|---|---|
| $\frac{1}{2}$ A — A′ = 50° — 5° 18′ 09″.36 = | 44° 41′ 50″.64 | sin. = | 9·8471791 |
| $\frac{1}{2}$ A = | 50° 00′ 00″.00 | co. ar. sin. = | 0·1157460 |
| G (ambiguous) | 66° 39′ 38″.22 | sin. = | 9·9629251 |
| G (corrected) | 113° 20′ 21″.78 | | |

Having found G, we proceed to find the radius $= r$. By formula (13′) and (14′) we have

$$b = \frac{\sin. (\frac{1}{2} A - A' + G)}{\sin. (\frac{1}{2} A - A)} \text{ and } r = \frac{bm}{b} = \frac{\sin. (\frac{1}{2} A - A').bm}{\sin. (\frac{1}{2} A - A' + G)}$$

| | | | | | |
|---|---|---|---|---|---|
| $\frac{1}{2}$ A — A′ | = 44° 41′ 50″·64 | | | | |
| G | = 113° 20′ 21″·78 | | | | |
| $\frac{1}{2}$ A — A′ + G | = 158° 02′ 12″·42 | .... | co. ar. .... | sin. = | 0·4271153 |
| $\frac{1}{2}$ A — A′ | = 44° 41′ 50″·64 | .... | .... | sin. = | 9·8471791 |
| $bm$ | = 2367·2198 feet | .... | .... | log. = | 3·3752386 |
| $r$ | = 4462·035 feet | .... | .... | log. = | 3·6495330 |

We thus find the radius of the curve, or $r = 4462 \cdot 035$ feet.

The deflection for a chord = 50 feet will be (3) sin. $D = \frac{\frac{1}{2} ch}{r}$

| | | | | |
|---|---|---|---|---|
| Thus, .... $\frac{1}{2}$ $ch$ | = 25 feet | | log. = | 1·3979400 |
| $r$ | = 4462·035 feet | co. ar. | log. = | 6·3504670 |
| D | = 0° 19′ 15″·67 | | sin. = | 7·7484070 |

But, as 0° 19′ 15″·67 makes an inconvenient number to add or subtract, we choose for the angle of deflection (D) = 0° 19′ 15″, and adopt a radius which shall agree therewith. This change in the radius will not materially alter the location of the curve.

To find a radius corresponding to a deflection of 0° 19′ 15″, we have (5) $r = \frac{\frac{1}{2} ch}{\text{sin. D}}$

| | | | | |
|---|---|---|---|---|
| $\frac{1}{2}$ *ch* | = 25 feet | | log. = | 1·3979400 |
| D | = 0° 19′ 15″ | .... co. ar. .... | sin. = | 2·2518454 |
| *r* | = 4464·63 feet | | log = | 3·6497854 |

We have thus ascertained the radius of a curve which will correspond to the location selected. It now remains to ascertain the tangent points, or points of commencement and end.

We have (6) representing the whole centre angle by C,

$$t = \tan.\ \tfrac{1}{2}\ C . r$$

| | | | | |
|---|---|---|---|---|
| Thus, .... | $\frac{1}{2}$ C | = 40° 00′ 00″ .... | tan. = | 9·9238135 |
| | *r* | = 4464·63 feet .... | log. = | 3 6497854 |
| | *t* | = 3746·27 feet .... | log. = | 3·5735989 |

We found (page 29) A T = 3536·98 feet, and A T′ = 1880·085 feet; and, *t* being 3746 · 27 feet, we have

*t* = 3746·27
A T = 3536·98

209·29 feet, the distance T should be moved from A

Again, *t* = 3746·27
A T′ = 1880·085

1866·185 feet, the distance T′ should be moved from A

Having moved the points T and T′ to their positions indicated above, and marked in the diagram T‴ and T″, the curve may be laid out and marked by any method the engineer might think best suited to the locality.

**(15)** We will add one more method of locating simple curves, principally applicable to large apex angles, and which may in some instances be practised beneficially with apex angles somewhat acute, provided the radii be not of great length.

This method supposes that the locality will admit of the lines to be connected by the curve, to be run or extended to their intersection, so that their apex angle may be measured directly; and that the contour of the surface is such that measurements may be taken with a good degree of accuracy, from the apex to the tangent points along these extended lines, and from these extended lines to the location of the curve.

Having premised thus much, let us suppose these extended lines, which we shall hereafter call the tangent lines, intersect each other at an angle of 170°, and it be desirable to connect these lines by a curve corresponding to a deflecting angle of 6′ for a chord of 50 feet. The elements required to be obtained from computation are,

First, Radius of the curve.

Second, Length of do.

Third, Distance from apex to tangent points.

Fourth, The number of primitive points convenience requires to be fixed in the curve.

To ascertain the radius, we have (5′) $r = \frac{\frac{1}{2} ch}{\sin. D}$

| | | | | |
|---|---|---|---|---|
| $\frac{1}{2}\ ch$ | = | 25 | | log. = 1·3979400 |
| D | = | 0° 6′ 00″ .... | co. ar. | .... sin. = 2·7581229 |
| Radius = $r$ | = | 14323·95 feet | | log. = 4·1560629 |

To find the length of the curve, we have (8) $a = \frac{r \,.\, C''}{r''}$

| | | | |
|---|---|---|---|
| | $r$ = | | log. = 4·1560629 |
| 180° — 170° | = C″ = 36,000″ ........ | | log. = 4·5563025 |
| | $r$ | co. ar. | log. = 4·6855749 |
| Length of arc = $a$ | = 2500·02 feet | | log. = 3·3979403 |

To find the distance from apex to tangent points, (6) $t = \tan. \frac{1}{2}\, C \,.\, r$

| | | |
|---|---|---|
| $\frac{1}{2}$ C = | 5° 00′ 00″ | tan. = 8·9419518 |
| $r$ | ..................... | log. = 4·1560629 |
| Apex to tan. = $t$ | = 1253·18 feet | log. = 3·0980147 |

We have found the length of arc = 2,500 feet; if we now suppose T to bear an even number in the locating stations, say 540, we may divide the arc into five equal parts of five hundred feet each, which will cause every point of division to fall on an even station in the location. This division, of course, divides the centre angle into five equal parts; and, as C = 180 — 170 = 10°, the centre angle corresponding to each division will be $\frac{10°}{5} = 2° \; 00' \; 00''$.

**(16)** Having determined to divide the curve into five equal parts, we now compute the distances from each of these dividing points in the curve to the tangent lines, in the direction of the radii passing through them. (See Fig. 5.)

Denoting the tangent points by T and T′; and the divisions of the curve by 1, 2, 3, 4; and the corresponding divisions of the tangent lines by *a b c d;* and representing by $C_1$ the centre angle corresponding to the arc T 1, and by $C_2$ the centre angle corresponding to the arc T 2. (The arcs T′ 4 and T′ 3, being similar to T 1 and T 2, will not need separate expressions.) In the general investiga-

[Fig. 5.]

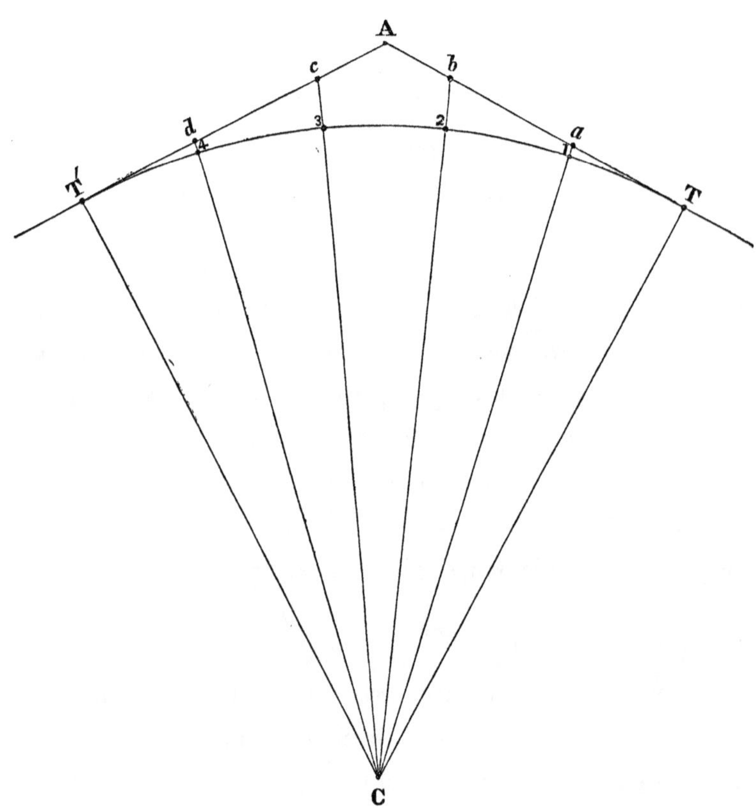

tions, we however denote the centre angles by C, and we have this analogy,

$$\text{Cos. C} : r :: \text{sin. C} : \text{T}\,a = \text{tan. C} . r \qquad (15)$$

$$\text{and Cos. C} : r :: \text{R} : \text{C}\,a \quad = \frac{r}{\text{cos. C}}$$

and we have $\frac{r}{\text{cos. C}} - r = a$, $1 =$ etc. (16) corresponding with the centre angle. Now, substituting $C_1$ and $C_2$ for C, as explained above, we have

| | | | | |
|---|---|---|---|---|
| $C_1$ | $= 2° 00' 00''$ | ........ | tan. = | 8·5430838 |
| $r$ | $= 14323·95$ feet | ........ | log. = | 4·1560629 |
| T′ to $d$ and T to $a$ = | 500·20 feet | ........ | log. = | 2·6991467 |
| $C_1$ | $= 2° 00' 00''$ co. ar. | ........ | cos. = | 0·0002646 |
| $r$ | $= 14323·95$ | ........ | log. = | 4·1560629 |
| C to $d$ and C to $a$ = | 14332·69 | ........ | log. = | 4·1563275 |
| ∴ $d$ to 4 and $a$ to 1 = | 8·74 feet | | | |
| Again, $C_2$ | $= 4° 00' 00''$ | ........ | tan. = | 8·8446437 |
| $r$ | = | ........ | log. = | 4·1560629 |
| T′ to $e$ and T to $b$ = | 1001·65 feet | ........ | log. = | 3·0007066 |
| $C_2$ | $= 4° 00' 00''$ co. ar. | ........ | cos. = | 0·0010592 |
| $r$ | $= 14323·95$ feet | ........ | log. = | 4·1560629 |
| C to $c$ and C to $b$ = | 14358·93 | ........ | log. = | 4·1571221 |
| ∴ $c$ to 3 and $b$ to 2 = | 34·98 feet | | | |

Having thus ascertained all the elements necessary to this peculiar method, we may now measure from T to $a$, in the direction T A = 500·2 feet; and the same distance from T′ to $d$, in the direction T′ A; and then, with the theodolite at $a$, and pointing to T, lay off the complement angle of $C_1$, and measure 8·74 feet to 1, for a point in the curve corresponding to station 540·00 + 5·00 = 545 of the location. Then, remove the theodolite to $d$, and pointing at T′, lay off the complement angle of $C_1$, and measure in the direction indicated 8·74 feet to 4, for a point in the curve corresponding to station (540·00 + 20·00) = 560 of the location. We now measure from *d* (1001·63 — 500·20) = 501·43 to *c*, and the same distance from *a* to *b*; then, with the theodolite at *b*, and pointing at T, lay off the complement angle of $C_2$, and measure in the direction indicated 24·98 feet to 2, for a point in the curve corresponding to station (540·00 + 10·00) = 550 of the location. Then, moving with the theodolite to *c*, and pointing to T′, lay off the complement angle of $C_2$, and measure in the direction indicated 24·98 to 3, for a point in the curve corresponding to station (540·00 + 15·00) = 555 of the location.

**(17)** We now take the theodolite successively to each of the points we have established in the curve; and, by deflections and corresponding chords, complete the work.

This method of laying out curves is found exceedingly convenient in woodlands, as not being so liable to mistakes which might lead the location astray as other methods, and will frequently save much trouble in chopping timber.

We are aware that the example we have given in the foregoing is one of the most convenient that the problem admits; but we think the principle will be sufficiently comprehended to apply it readily and without difficulty in its most complicated form, without further explanation.

**(18)** We have now completed all we contemplated respecting the investigations of the practical operations of laying out simple railroad curves. It was not our purpose to pursue these investigations until the subject was exhausted. That would have taken a long time, and might have occupied much room, and it is more than probable that it would have exceeded our ingenuity and ability. But we hope we have said enough to give the proper direction to the inquiries of the young or inexperienced engineer, and to convince him of the necessity of making the study of the elements of geometry and trigonometry (if it be proper to make the distinction, or to class them under different heads) a matter of the first importance.

**(19)** We now proceed to the consideration of reverse curves; the most simple form of which is to unite two parallel lines, which, continued, will not intersect or run into each other, by curves of equal radii.

This problem is of such simplicity as to admit of many forms of construction, and great variety of formulæ; and, had we not come to the conclusion to give a formula for laying out curves to unite railroad lines under every condition which has occurred in our practice, we think we should have passed by this problem without considering its properties. We therefore content ourselves with giving the following rules and formulæ.

**(20)** Let T A and T′ B represent two lines of railroad having the same bearing, but so located that they will not intersect by extension.

Let T C and T′ C′ = the radii which we suppose to be equal in length, and which in the investigation we shall represent by $r$.

[Fig. 6.]

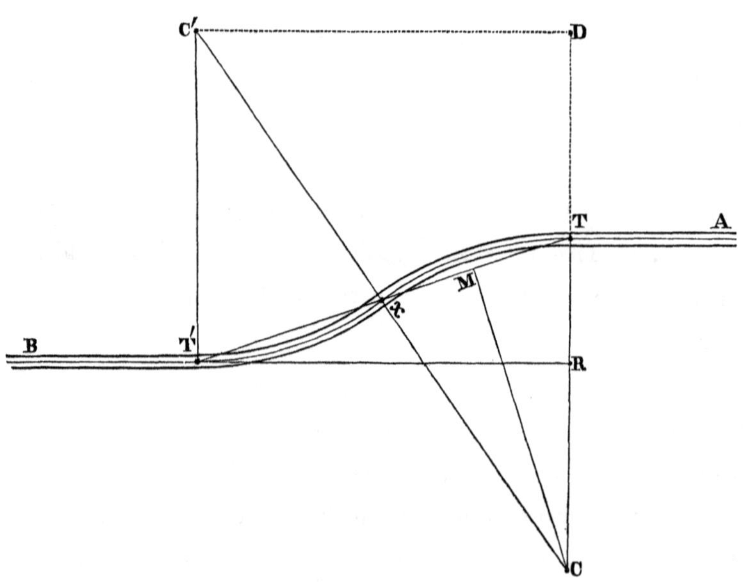

Draw the line T C and T′ C′ at right angles with T A and T′ B, and make each $= r$; then, drawing a line from C to C′ will intersect the curves at their reversing point X, (see figure,) so also a line drawn from T to T′ will intersect the curves at the same point, and will also bisect the line C C′.

We now extend the line B T′ until it intersects the line or radius T C at R, and as T C is drawn at right angles with T A, and T A and B T′ having the same bearing, it is obvious that the angle at the intersection will be a right angle. Then, having measured T′ R and T R, or ascertained their length by computation, we have by considering the line = T T′ as radius, and the line T′ R as a cosine, and the line R T as a sine of the angle T T′ R, this analogy to find the angle T′. Representing the sine by $s$, the cosine by $c$, the radius by R, then will

$$c : s :: R : \tan. T' = \frac{s}{c} \qquad (17)$$

By letting fall a perpendicular upon the line T X from C, we divide the triangle T C X into two equal right-angled triangles, viz., C M T and C M X; now, as the triangles T T′ R and T C M have the angle T common to both, it is obvious they are similar; the angles T′ and C must therefore be equal, and in the triangle T C X the angle C will be equal to twice T′; then, resolving the triangle T′ R T, we find the length T T′, half of which is equal to T X; then solving the triangle C X T, we find the side T C = radius $= r$. To execute these computations by the aid of the trigonometrical functions, we have in the triangle T T′ R to find ½ T T′ which we represent by $a$.

Sin. T′ : $s$ :: R : T T′; wherefore $\frac{T\,T'}{2} = a = \frac{s}{\sin. T' . 2}$

or, Cos. T′ : $c$ :: R : T T′; " $\frac{T\,T'}{2} = a = \frac{c}{\cos. T' . 2}$

Then, in the triangle T C X we have

$$\text{Sin. C} : a :: \text{cos. T}' \cdot r = \frac{a \cdot \text{cos. T}'}{\text{sin. C}} = \frac{c}{\text{sin. C} \cdot 2} \quad (18)$$

Having thus found the radius, and the centre angle C of the one half the curve, it is evident that the other half must contain identical elements.

**(21)** The curve may be laid out in accordance with such of the formulæ described in the foregoing pages, relating to simple curves, as the engineer may think best suited to the condition of things and the contour of the surface.

For the purpose of presenting an example of computation, we will suppose

$$\text{T}' \text{R} = c = 1260 \text{ feet}$$

$$\text{R T} = s = 150 \text{ ''}$$

Then, by (17) $\quad \text{Tan. T}' = \frac{s}{c}$

| | | | |
|---|---|---|---|
| Wherefore $s$ = 150 feet | | ........ log. = | 2·1760913 |
| $c$ = 1260 " | co. ar. | ........ log. = | 6·8996295 |
| T′ = 6° 47′ 20″·31 | | ........ tan. = | 9·0757208 |

C being equal to 2 T, we have (18) $r = \frac{c}{\text{sin. C} \cdot 2}$

Wherefore T′ = $\frac{6° \ 47' \ 20''{\cdot}31}{2}$

| | | | |
|---|---|---|---|
| C = 13° 34′ 40″·63 | co. ar. | ........ sin. = | 0·6293609 |
| 2 | " " | ........ log. = | 9·6989700 |
| $c$ = 1260 | | ........ log. = | 3·1003705 |
| $r$ = 2683·47 | | ........ log. = | 3·4287014 |

**(22)** We will now vary the given elements, by supposing the radius $= r = 2683 \cdot 47$ feet, and the distance between the centres of the tracks $= s = 150$, to find the relative positions of T′ and T. This case supposes a beginning point to have been selected. Suppose that point to be at T. Then, from T we draw the radius T C $= r$ at right angles to the line T A, and from T we extend the line T C or $r$ in the opposite direction, a distance $= r$ — (or minus) the distance between the tracks (which we have supposed to be 150 feet) to D; and from D draw the line D C′ at right angles with the line D C; then, with the line C C′, equal in length to $2\,r$, intersect the line D C′ at C′; then will D C′ = T′ R, representing T′ R as in our former proposition by $c$; and then, solving the triangle C C′ D, we have by trigonometry,*

$$2\,r : \text{R} :: (2\,r - s) : \sin. \text{C}' = \frac{(2\,r - s)}{2\,r} \qquad (19)$$

and $$\text{Sin. C}' : (2\,r - s) :: \cos. \text{C}' : c = \cot. \text{C}'\,(2\,r - s) \qquad (20)$$

The complement of the angle C′ found by the above formula = the angles T C C′ = T′ C′ C.

EXAMPLE OF COMPUTATION.

By formula (19)

| | | | |
|---|---|---|---|
| $2\,r$ | = 5367 co. ar. | ........ log. | = 6·2702685 |
| $(2\,r - s)$ | = 5217 | ........ log. | = 3·7174298 |
| C′ | = 76° 25′ 19″.50 | ........ sin. | = 9·9876893 |

By formula (20)

| | | | |
|---|---|---|---|
| C′ | = 76° 25′ 19″.50 | ........ cot. | = 9·3829486 |
| $(2\,r - s)$ | = | ........ log. | = 3·7174208 |
| $c$ | = 1260 nearly | ........ log. | = 3·1003694 |

* These formulæ are generally based on trigonometry, as being more convenient or better suited to logarithms.

The complements to C′ as found above = 90 — 76° 25′ 19″.5 = 13° 34′ 40″.5.

**(23)** There being a difference of opinion among practical engineers respecting the propriety of reversing curves without the intervention of a piece of straight track between them, we have thought it would not be improper, before we proceed further with our investigations of formulæ for reverse curves, to add the following discussion of the causes of the lateral shocks experienced in railroad cars when entering upon a curve; and also the necessity for a piece of straight track between reversing curves.

From any investigations that we have been enabled to make, we cannot discover any reason why a car should receive a lateral shock when running off a straight track on to a curve, provided the distance between the flanges of the car wheels and the distance between the insides of the rails are the same; and the outside rail of the curve be properly elevated to suit the velocity. We would however remark, if the curve be very short and the velocity rapid, the friction of the flanges of the wheels upon the outside rails of the curve will become so great as to occasionally raise the face or bearing of the wheel from the rail, which falls again suddenly upon the track, producing a perpendicular jar or concussion, which may sometimes occasion lateral motion; but, if the curves are in good order, and the rails properly curved, this motion will be seldom felt, unless, I repeat, the radius be very short, and the velocity great. In this connection I would add, it is believed to be somewhat dangerous to run a train over a curve in proper repair or condition with a velocity so great as to occasion this phenomenon. We have, however, felt this motion when riding on curves not in proper repair, or where the rails have not been curved in a uniform, regular manner; whereas, if the track had been in good condition, nothing of the kind would have been felt.

Neither can we, from any investigation we have been enabled to make, discover any reason for a lateral shock to a car when running off of a direct on to a reverse curve (if these are the proper terms of expression) provided the condition of the track is good, and the wheel flanges and the rails are the same distance apart. But the question may be asked: Why in practice lateral shocks are so frequently felt, when running from a straight line on to a curve; and likewise when running from a direct on to a reverse curve? We answer, that the practice is almost universal to lay the rails of a track from one half to three fourths of an inch further apart than the flanges of the car wheels. Now if we conceive the flanges of the car wheels to be in contact with the line of rail which forms the inside line of the curve, when the car is about to enter upon that curve, and the track three fourths of an inch wider than the flanges, it will be obvious that the motion of the car, if it meets with no extraordinary obstruction, will continue straight, or in a direct line, after it enters the curve, until the flange of the wheel meets with the outside rail of the curve; the distance which the car will have then advanced into the curve, before the phenomenon of contact takes place, will be so great that the curve will have obtained a considerable degree of deflection, which of course produces a shock, and the shock will be somewhat proportioned to the difference between the width of the flanges of the wheels and the width of the track, the velocity of the cars, and the length of the radius of the curve. The following diagram (Fig. 7) indicates the practice in cases of reversed curves.

[Fig. 7.]

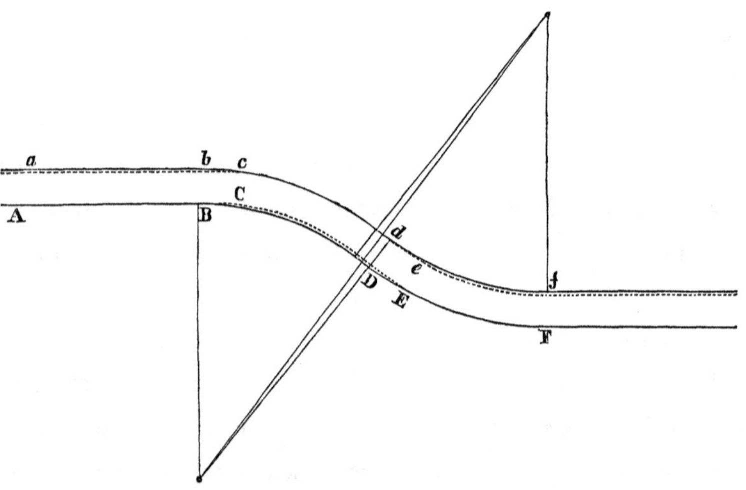

(**24**) Let us now examine the condition of a car running off of a direct curve on to a reverse. The centrifugal force, as well as the position of the wheel axles, direct a car when running on a curve, to the outside rail, which must, of course, become its guide; and when the car arrives at the reversing point, or rather one fourth of its length between its wheels beyond that point, it will continue its direction in a straight line until its wheel flanges meet with the outside rail of the reverse curve, and if the difference of width between the flanges of its wheels and the rails be considerable, the car will have advanced so far into the curve before the phenomena of contact take place, as to admit the rail taking a considerable degree of deflection; and, as before stated respecting simple curves, the meeting of the wheel flanges with the deflected rail causes a lateral shock which is sensibly felt, and is proportioned to the velocity of the cars, the difference in width between the flanges of the wheels and the rails, and the length of the radius of the curve.

The motion of the cars from the direct to the reverse curve will always be of the same character under similar circumstances; the cars being constantly influenced by the position of their axles, and the centrifugal force of motion. But it is not so with the cars upon a straight track; there is nothing to uniformly guide them to the side of the track which forms the inner rail of the curve when running off of a straight line on to a curve; and, if the flanges of the wheels are in contact with the side of the track forming the outer rail of the curve, the car will enter upon the curve without lateral shock.

were laid down to correspond with the flanges of the wheels, a car will meet with no greater lateral shock when running through reversing points of a reverse curve, than when running off of a straight track on to a simple curve; hence, under restricted circumstances, where reversed curves are required, of short radius, it becomes an object of importance for the curves to occupy the whole line, that is, there should be no straight track between them; as a straight line of any considerable length will tend to diminish the length of the radius.

**(25)** Let us now endeavor to explain this matter by diagrams. We will suppose a car to be running on a straight track from A towards F, with the flanges of the wheels in contact with the rail A B, the flanges of the wheels on the opposite side will not be in contact with the bar *a b*, but will describe the dotted line parallel with it; the car arrives at the tangent point B *b*, its natural motion will be from *b* B to *c* C, in the direction of the dotted lines, where the flange of the wheel running upon the rail *a f* meets it at the point c, some distance advanced upon the curve, the rail at this point making an angle with the direction of the car, causes a shock and a sudden lateral motion; the car then proceeds onwards, with the flanges in contact with the rail *a f*, until it arrives at the point D *d;* from B to D the flanges have not been in contact with the rail A F, but have described the dotted line parallel with it; from D *d* its motion is onward in the direction of the tangent until it arrives at E *e*, when the flange meeting with the rail at E, which it will be seen forms an angle with the direction of the car, causes the lateral shock felt at this point; the car then moves onwards, with the flanges in contact with the rail A F, until it passes through or over the curve, and no further shock is felt. In the mean time the flanges upon the opposite side describe the dotted line parallel with *e f.* If, therefore, our reasoning be correct, it will be obvious that if we would have cars run smoothly over a railroad, the track should, near the tangent points of curves, be laid down to correspond with the width between the flanges of the wheels; and we add that the same thing should be observed near the turn-out frogs, as it is important that the scores in the frogs through which the flanges of the wheels traverse, should be just sufficient to permit them to pass. To render this practice complete calls for a greater degree of care in the adjustment of wheels upon their axles, than is at present practised in many constructing and repair shops; but, in the present careless condition of adjustment, the management of tracks can be much improved. Near frogs, and the commencement of curves, the rails of the track should be no wider than the widest wheels of the train. A further improvement, adapted to the passage through the score in the frog smoothly, is to have the width between the back side of the wheel flanges as near alike as they well can be, which will much improve the benefits of the guard rails.

In this connection it may not be amiss to compute the deflections of the rail at the point where it is met by the flanges of the wheels when running into the curve; assuming the rails of the track from 0·01 of a foot wider than the flanges, up to 0·06 of a foot; and supposing the car to be tracing the inner rail on its approach to the tangent point, and the curve to be of a radius of 1000 feet.

The following investigation, and its examples of computation, by referring to Fig. 6, may enable the student to master the whole merits of this subject.

## INVESTIGATION OF FORMULA, AND EXAMPLES OF COMPUTATION.

Let $r$ = radius of the curve; $h$ the width between the rails of the track;

$\triangle_1, \triangle_2, \triangle_3$, etc., the differences of distance between the tracks and the flanges of the wheels;

$c_1, c_2, c_3$, etc., the corresponding angles of deflection.

Then, putting $r = 1000$ feet; $h = 4{\cdot}7$ feet; $\triangle_1, \triangle_2, \triangle_3$, etc. $= 0{\cdot}01, 0{\cdot}02, 0{\cdot}03$ feet, etc., we have

$$r + \tfrac{1}{2} h : \mathrm{R} :: r + \tfrac{1}{2} h - \triangle_1, \text{ etc.} : \cos. c = \frac{r + \frac{1}{2} h - \triangle_1 \text{ etc.}}{r + \frac{1}{2} h}$$

| | | | | |
|---|---|---|---|---|
| Thus, | $r + \frac{1}{2} h$ | = 1002·35 co. ar. | ........ | log. = 6·9989806 |
| | $r + \frac{1}{2} h - \triangle_1$ | = 1002·34 | ........ | log. = 3·0010150 |
| | $c_1$ | = 0° 15′ 26″ | ........ | cos. = 9·9999956 |
| | $r + \frac{1}{2} h$ | = 1002·35 co. ar. | ........ | log. = 6·9989806 |
| | $r + \frac{1}{2} h - \triangle_2$ | = 1002·33 | ........ | log. = 3·0010107 |
| | $c_2$ | = 0° 21′ 44″ | ........ | cos. = 9·9999913 |
| | $r + \frac{1}{2} h$ | = 1002·35 co. ar. | ........ | log. = 6·9989806 |
| | $r + \frac{1}{2} h - \triangle_3$ | = 1002·32 | ........ | log. = 3·0010064 |
| | $c_3$ | = 0° 26′ 34″ | ........ | cos. = 9·9999870 |
| | $r + \frac{1}{2} h$ | = 1002·35 co. ar. | ........ | log. = 6·9989806 |
| | $r + \frac{1}{2} h - \triangle_4$ | = 1002·31 | ........ | log. = 3·0010020 |
| | $c_4$ | = 0° 30′ 45″ | ........ | cos. = 9·9999826 |
| | $r + \frac{1}{2} h$ | = 1002·35 co. ar. | ........ | log. = 6·9989806 |
| | $r + \frac{1}{2} h - \triangle_5$ | = 1002·30 | ........ | log. = 3·0009977 |
| | $c_5$ | = 0° 34′ 20″ | ........ | cos. = 9·9999783 |
| | $r + \frac{1}{2} h$ | = 1002·35 co. ar. | ........ | log. = 6·9989806 |
| | $r + \frac{1}{2} h - \triangle_6$ | = 1002·29 | ........ | log. = 3·0009934 |
| | $c_6$ | = 0° 37′ 35″ | ........ | cos. = 9·9999740 |

The foregoing computed deflections of the track, at the point met by the wheel flanges, under the assumed circumstances, show the necessity of narrowing the guage near the commencement of curves, and near the reversing points in reverse curves. There can, however, be no doubt but the cars will run steadier and safer over a narrow track, just suiting the wheel flanges, than over one of greater width. The only argument in favor of the guage of a track being wider than the flanges of the wheels is, that a greater surface of the wheel is exposed to wear upon the rails by the zig-zag course which the wide guage allows the cars to take, than would be if that motion was prevented.

**(26)** The next form of reverse curves which we shall consider, is that which shall unite two lines having different bearings, which of course would intersect each other were they continued; but, on account of avoiding some obstacles, or the desire of a near approach to some particular locality, it becomes necessary to connect these lines by reverse curves; it is, therefore, a matter of great importance to lay down these curves in the best form possible, particularly if they require short radii, which will be best accomplished, if there be no obstacle in the way, by making the curves of equal curvature, and occupying the whole distance between the tangent points; (these tangent points are supposed to be fixed by the contour of the surface, or some other consideration or governing principle, which cannot well be avoided.)

**(27)** To proceed with the investigation of the proper formula for determining the elements of these curves, we would first remark, that this problem requires the angles A T′ T and T′ T B, and also the length of the line T T′ to be measured. (See Fig. 8.)

[Fig. 8.]

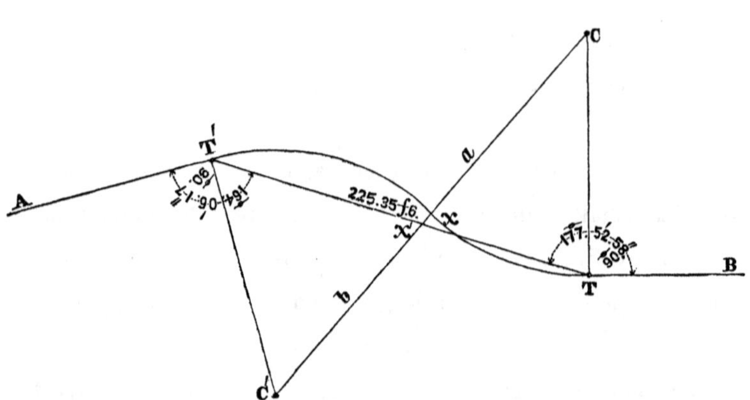

Then, by putting T′ X′ = $c'$, T X = $c$, C X = $a$, C′ X = $b$, and T′ T as measured = $m$, and the radii C T and C′ T′ in proportion = unity; (these radii by the problem being equal,) and the radius in measure = $r$.

Then, by the problem we have the line C′ C = twice the length of radius, and it will be apparent by a glance at the diagram that the angles X and X′ must be equal.

Now, commencing with the radius = unity, we have

$$\text{Sin. X} : 1 :: \text{sin. } (T - 90^\circ) : a = \frac{\text{sin. } (T - 90^\circ)}{\text{sin. X}}$$

$$\text{Sin. X}' : 1 :: \text{sin. } (T' - 90^\circ) : b = \frac{\text{sin. } (T' - 90^\circ)}{\text{sin. X}'}$$

Substituting for $a + b$ their value, viz., twice radius, and as we have taken radius = unity, we have this equation,

$$\frac{\text{Sin. } (T - 90^\circ)}{\text{sin. X}} + \frac{\text{sin. } (T' - 90^\circ)}{\text{sin. X}} = 2$$

Multiplying by sin. X, we have

$$\text{Sin. } (T - 90^\circ) + \text{sin. } (T' - 90^\circ) = 2 \text{ sin. X; hence}$$

$$\text{Sin. X} = \frac{\text{sin. } (T - 90^\circ) + \text{sin. } (T' - 90^\circ)}{2} \qquad (21)$$

Having found the angle X, we next have

$180^\circ - (T - 90^\circ) - X = C$; and $180^\circ - (T' - 90^\circ) - X = C'$

Then, $\quad \text{Sin. X} : 1 :: \text{sin. C} : c = \frac{\text{sin. C}}{\text{sin. X}} \qquad (22)$

$\quad \text{Sin. X} : 1 :: \text{sin. C}' : c' = \frac{\text{sin. C}'}{\text{sin. X}} \qquad (23)$

And, $\quad c + c' : m :: 1 : r = \frac{m}{c + c'} \qquad (24)$

Having thus obtained the radius, and the angles required, the remaining elements necessary for making or laying out the curve may, of course, be computed by such of the foregoing formulæ as the condition of the locality requires.

**(28)** NOTE. Because of the scarcity of extensive tables of natural sines, and for the purpose of showing how readily they can be obtained from logarithmic sines, we have, in the specimens of computation given below, obtained the natural sine from the logarithmic sine, and after having found the natural sine of X, we have deduced its logarithm, and then ascertained the corresponding angle from the tables of logarithmic sines.

We have deemed it proper to give the above hints, for the information of such young engineers as may not be familiar with the principles of trigonometrical tables.

To proceed with the examples of computation, we have

| | | | |
|---|---|---|---|
| T — 90° = 87° 52′ 58″ | log. sin. = 9·9997034 | nat. sin. = | 0·9993172 |
| T′ — 90° = 74° 06′ 17″ | log. sin. = 9·9830685 | nat. sin. = | 0·9617640 |
| Nat. sin. (T — 90°) + nat. sin. (T′ — 90°) | | = | 2)1·9610812 |
| X = 78° 40′ 42″ | log. sin. = 9 9944656 | nat. sin. = | 0·9805406 |

| | | | |
|---|---|---|---|
| We have found X | = 78° 40′ 42″ | Again, X | = 78° 40′ 42″ |
| and (T — 90°) | = 87° 52′ 58″ | and (T′ — 90°) | = 74° 06′ 17″ |
| Wherefore C must | = 13° 26′ 20″ | ∴ C must be | = 27° 13′ 01″ |
| Proof | 180° 00′ 00″ | Proof | 180° 00′ 00″ |

| | | | |
|---|---|---|---|
| By (22) we now have | X = 78° 40′ 42″ co. ar. | ........ sin. = | 0·0085344 |
| | C = 13° 26′ 20″ | ........ sin. = | 9·3662513 |
| | $c$ = 0·2370204 | ........ log. = | 9·3747857 |
| By (23) = again | X = 78° 40′ 42″ co. ar. | ........ sin. = | 0·0085344 |
| | C′ = 27° 13′ 01″ | ........ sin. = | 9·6602591 |
| | $c'$ = 0·4664375 | ........ log. = | 9·6687935 |
| By (24) | $c + c'$ = 0·7034579 co. ar. | ........ log. = | 0·1527619 |
| | $m$ = 225·35 feet | ........ log. = | 2·3528576 |
| | $r$ = 320·346 " | ........ log. = | 2·5056195 |

Having ascertained the angles and the radius, we do not think it necessary to extend further the examples of calculation for this particular case.

**(29)** We sometimes have another form of the reverse curve, which we will endeavor to investigate. It sometimes happens that in a condition of the tangent lines not very unlike the last described, we have some particular points which we are desirous should govern the location of the track. This state of things necessarily fixes the length of one of the radii, and it is our object in the investigation, to deduce a formula for ascertaining the length of the other, with the centre angles which measure the arcs, etc.; it being apparent that fixing the length of the radius of one of the curves, governs the radius of the other.

To proceed with the investigation. Let A T′ and T B represent the tangent lines, (see figure 9;) T′ and T the measured angles, viz., A T′ T and B T T′; and T′ T the line measured, which we represent by $m$; then,

Putting $a$ for the line C′ E = R T or rather = D T + R D
" $b$ " " E T = C′ R
" $r$ = the given radius
" $x$ = the radius required
" $c'$ = the line R D

we have $a^2 + (\overline{b + x})^2 = (\overline{r + x})^2$

Expanding the equation, $a^2 + b^2 + 2\,bx + x^2 = r^2 + 2\,rx + x^2$

Subtracting $x^2$ leaves $a^2 + b^2 + 2\,bx = r^2 + 2\,rx$

Trans. and changing signs $a^2 + b^2 - r^2 = 2\,rx - 2\,bx$

Dividing by 2 $\frac{a^2 + b^2 - r^2}{2} = (r - b)\,x$

and by $(r - b)$ $\frac{a^2 + b^2 - r^2}{2\,(r - b)} = x$ (25)

[Fig. 9.]

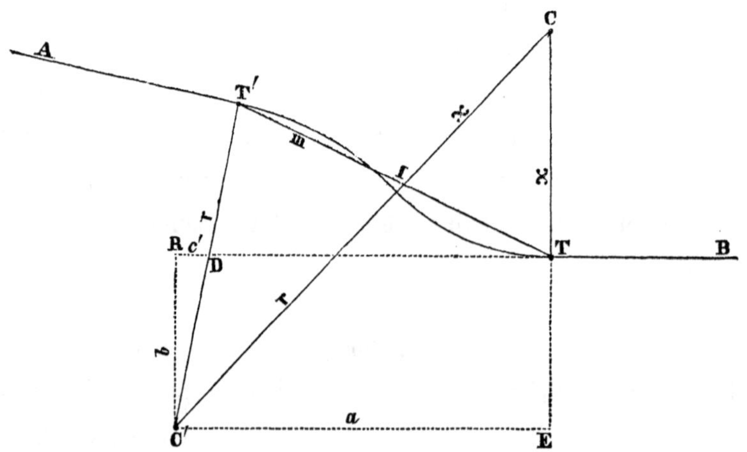

Having deduced the formula, we proceed to give a specimen of calculation. We will suppose $m = 630$ feet

$r = 555$ "

$T' = 169° 30'$

$T = 157° 00'$

To ascertain $a$ we have first to find T D; in the triangle T T′ D we have the

| | | | | | |
|---|---|---|---|---|---|
| Angle T = (180° — 157°) = | 23° 00′ 00″ | D | = 77° 30′ | co. ar. sin. = | 0·0104185 |
| T′ = (169·30 — 90°) = | 79° 30′ 00″ | $m$ | = 630 | log. = | 2·7993405 |
| D = Supplement = | 77° 30′ 00″ | T′ | = 79° 30′ | sin. = | 9·9926661 |
| Proof = | 180° 00′ 00″ | T D | = 634·4905 feet | log. = | 2·8024251 |

| | | | | | |
|---|---|---|---|---|---|
| | $\frac{m}{\text{Sin. D}}$ | | | log. = | 2·8097590 |
| | T | = | 23° 00′ | sin. = | 9·5918780 |
| | T′ D | = | 252·1373 | = | 2·4016370 |
| | $r$ | = | 555·0000 | | |
| In triangle C D R, $r$ — T′ D = C D | | = | 302·8627 | log. = | 2·4812458 |
| | D | = | 77·30 | sin. = | 9·9895815 |
| | $b$ | = | 295·6837 | log. = | 2·4708273 |
| | D | = | 77° 30′ | cot. = | 9·3457552 |
| | $c'$ | = | 65·5515 | log. = | 1·8165825 |
| From above | T D | = | 634·4905 | | |
| (T D + $c'$) = | $a$ | = | 700·0420 | log. = | 2·8451241 |
| | | | | | 2 |
| | $a^2$ | = | 490058·8 feet | log. = | 5·6902482 |
| | $b$ | = | 295·6837 | log. = | 2·4708273 |
| | | | | | 2 |
| | $b^2$ | = | 87428·82 feet | log. = | 4·9416546 |
| | $r$ | = | 555·00 | log. = | 2·7442930 |
| | | | | | 2 |
| | $r^2$ | = | 308025·00 feet | log. = | 5·4885860 |

[Fig. 10.]

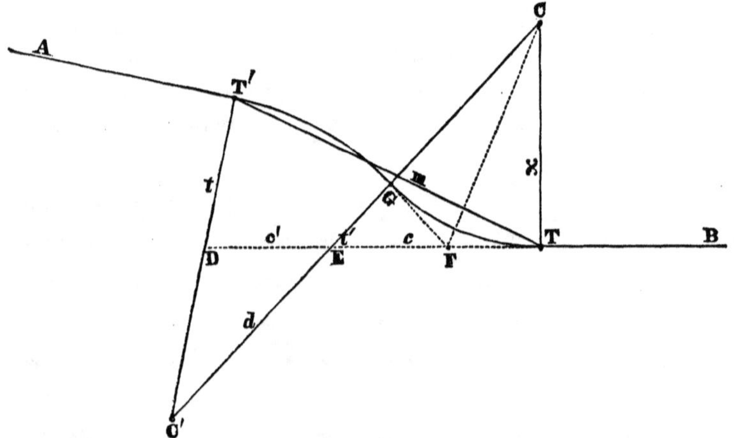

| | | | |
|---|---|---|---|
| $a^2 + b^2 - r^2$ | $=$ 269462·62 | log. $=$ | 5·4304985 |
| $2\ (r - b)$ | $=$ 518·6326 co. ar. | log. $=$ | 7·2851402 |
| $x$ | $=$ 519·5636 feet | log. $=$ | 2·7156387 |
| $b$ | $=$ 295·6837 | | |
| $x + b =$ E C | $=$ 815·2473* co. ar. | log. $=$ | 7·0887106 |
| $a$ | $=$ 700·0420 | log. $=$ | 2·8451241 |
| C | $=$ 40° 39′ 08″·09 | tan. $=$ | 9·9338347 |
| T′ ∽ T | $=$ 12° 30′ 00″·00 | | |
| C′ | $=$ 28° 09′ 08″·10 | | |

Having found the centre angle C, we can readily, by several methods, ascertain the value of the other centre angle, C′; we however shall only give the following method, viz.:

When the angle D in the triangle T′ D T, is greater than a right angle, the difference between the angles A T′ T and B T T′ must be added to C, and the sum will be equal to C′; and when D is smaller than a right angle, it must be subtracted.

The remainder of the elements, which may be needed to facilitate the operations of location, may without difficulty be found, by such of the foregoing formulæ as shall be found applicable.

**(30)** Another form of the reverse curve is wherein we may have one tangent point fixed, and one centre angle given, the tangent lines being located in position and direction, but one of them may be shortened or lengthened to adapt it to the unknown or required angles. This case occurs when the point where the curves reverse becomes a governing point in the location, as at the point G in the figure.

* Having in the triangle C′ C E found the sides, C′ E $= a$, and C E $= x + b$, we have to find the angle C; $\frac{a}{x + b} =$ tan. C.

## INVESTIGATION OF FORMULÆ.

In the diagram the tangent lines are represented by A T′ and B T; and the angles A T′ T and B T T′ by $T'_m$ and $T_m$; the line measured, viz., T′ T by *m*.

| | | | | | |
|---|---|---|---|---|---|
| Then, putting | $t'$ | for | the | line | D T |
| And, | $t$ | " | " | " | D T′ |
| | $r$ | " | " | given radius | C′ T′ |
| | $r'$ | " | " | radius sought | G C |

We have in the solution of this problem the following triangles, viz., T′ D T, which for convenience we denominate No. 1; C′ D E, No. 2; E F G, No. 3; F G C, No. 4; which require to be successively solved.

Commencing with triangle No. 1, we have

The angle at T $= 180° - T_m$

" " " T′ $= T'_m - 90°$

" " " D = the supplement of the above.

Then, by analogy, $\text{Sin. D} : m :: \text{sin. T} : t = \frac{m \text{ sin. T}}{\text{sin. D}}$ (26)

And, $\text{Sin. D} : m :: \text{sin. T}' : t' = \frac{m \text{ sin. T}'}{\text{sin. D}}$ (27)

In the solution of triangle No. 2, we have the side C′ D $= r - t$, and the angle D = the supplement of D in triangle No. 1; the angle C′ being given, the angle E = the supplement of C′ + D. Denoting the side C E by *d*, and the side D E by $c'$, we have

$$\text{Sin. E} : r - t :: \text{sin. D} : d \quad (28)$$

$$\text{and Sin. E} : r - t :: \text{sin. C}' : c' \quad (29)$$

In triangle No. 3 we have $r - d =$ E G; the angle E the same as E in No. 2; the angle G a right angle; and of course the angle F becomes the complement of E. Denoting the side G F by *e*, we have $\text{Cos. E} : r - d :: \text{sin. E} : e = \text{tan. E} (r - d)$

Then, in the quadrilateral F G C T, we have the angle at F = (180° — F), F being the same as in No. 3; bisecting F as thus found by a line from F to C, we have in triangle No. 4,

$$\text{Cos. } \tfrac{1}{2} \text{ F} : e :: \text{sin. } \tfrac{1}{2} \text{ F} : r' = \text{tan. } \tfrac{1}{2} \text{ F} . e = \text{tan. } \tfrac{1}{2} \text{ F} . \text{tan. E} . (r - d) \qquad (30)$$

The last expression being equal to the side G C.

In practice it will be convenient to ascertain the distance of the tangent point from the point T, and for that purpose we have in the triangle C T E, which we call No. 5,

$$\text{Sin. E} : r' :: \text{cos. E} : c_1 = \text{cot. E} . r' \qquad (31)$$

where $c_1$ represents the line E T.

Then, will $(c' + c_1) \backsim t' =$ the distance of the tangent from the point T, and the direction of course will be known from the relative magnitude of the numbers represented by $(c_1 + c')$ and $t'$, viz., if $t'$ represent the larger number, the tangent point will be in the direction of D; if the smaller, in the direction of B.

### EXAMPLE OF COMPUTATION.

Suppose $m = 630$ feet; $r = 555$ feet; $T'_m = 169° \ 30'$; $T_m = 157° \ 00'$; $C' = 27° \ 49' \ 27''\cdot35$.

Then, $T = (180° - T_m) = 180° - 157° = 23° \ 00'$

$T' = (T'_m - 90°) = 169° \ 30' - 90° = 79° \ 30'$

$D =$ supplement $77° \ 30'$

| By formula (26) | | | | |
|---|---|---|---|---|
| D | = 77° 30′ | co. ar. | sin. | = 0·0104185 |
| $m$ | = 630 feet | | log. | = 2·7993405 |
| T | = 23° | | sin. | = 9·5918780 |
| $t$ | = 252·1373 | | log. | = 2·4016370 |

| | | | | |
|---|---|---|---|---|
| (27) D | = 77° 30′ | co. ar. | sin. | = 0·0104185 |
| $m$ | = 630 feet | | log. | = 2·7993405 |
| T′ | = 79° 30′ | | sin. | = 9·9926661 |
| $t'$ | = 634·4905 feet | | log. | = 2·8024251 |
| (28) E | = 49° 40′ 32″·65 | co. ar. | sin. | = 0·1178203 |
| 555 — 252·1373 = $r - t$ | = 302·8627 feet | | log. | = 2·4812458 |
| D | = 102° 30′ | | sin. | = 9·9895815 |
| $d$ | = 387·836 feet | | log. | = 2·5886476 |
| (29) E | = 49° 40′ 32″·65 | co. ar. | sin. | = 0·1178203 |
| $r - t$ | = 302·8627 feet | | log. | = 2·4812458 |
| C′ | = 27° 49′ 27″·35 | | sin. | = 9·6690948 |
| $c'$ | = 185·4219 feet | | log. | = 2·2681609 |
| (30) ½ F | = 69° 50′ 16″·32 | | tan. | = 0·4351232 |
| E | = 49° 40′ 32″·65 | | tan. | = 0·0711997 |
| $r - d$ | = 167·164 feet | | log. | = 2·2231428 |
| $r'$ | = 536·3715 feet | | log. | = 2·7294657 |
| (31) E | = 49° 40′ 32″·65 | | cot. | = 9·9288003 |
| $r'$ | = | | log. | = 2·7294657 |
| $c'$ | = 455·267 feet | | log. | = 2·6582660 |
| $c_1$ | = 185·422 | | | |
| $c' + c_1$ | = 640·689 | | | |
| $t$ | = 634·490 | | | |
| $(c_1 + c') \backsim t$ = | 6·199 feet = | | | |

the distance the tangent point must be fixed from T towards B.

Again, suppose $m = 630$ feet; $r = 555$ feet; $T'_m = 169° \ 30'$; $T_m = 157°$; $C' = 28° \ 09' \ 08''·1$.

Then, T = 160° — 157° = 23° 00′

T′ = 169° 30′ — 90° = 79° 30′

D = supplement = 77° 30′

180° 00′

| | | | | | |
|---|---|---|---|---|---|
| By formula (26) | D | = 77° 30′ | co. ar. | sin. | = 0·0104185 |
| | $m$ | = 630 feet | | log. | = 2·7993405 |
| | T | = 23° 00′ | | sin. | = 9·5918780 |
| | $t$ | = 252·1373 feet | | log. | = 2·4016370 |
| (27) | D | = 77° 30′ | co. ar. | sin. | = 0·0104185 |
| | $m$ | = 630 feet | | log. | = 2·7993405 |
| | T′ | = 79° 30′ | | sin. | = 9·9926661 |
| | $t'$ | = 634·4905 feet | | log. | = 2·8024251 |
| (28) | E | = 49° 20′ 51″·9 | co. ar. | sin. | = 0·1199428 |
| | $r - t$ | = 302·8627 feet | | log. | = 2·4812458 |
| | D | = 102° 30′ | | sin. | = 9·9895815 |
| | $d$ | = 389·7356 feet | | log. | = 2·5907701 |
| (29) | E | = 49° 29′ 51″·9 | co. ar. | sin. | = 0·1199428 |
| | $r - t$ | = 302·8627 feet | | log. | = 2·4812458 |
| | C′ | = 28° 09′ 08″·1 | | sin. | = 9·6737728 |
| | $c'$ | = 188·3482 feet | | log. | = 2·2749614 |
| (30) | ½ F | = 69° 40′ 25″·95 | | tan. | = 0·4312942 |
| | E | = 49° 20′ 51″·90 | | tan. | = 0·0661653 |
| $r = 555$; $d = 389·7356$, | $r - d$ | = 165° 26′ 44″ | | log. | = 2·2181792 |
| | $r$ | = 519·5636 feet | | log. | = 2·7156387 |
| (31) | E | = 49° 20′ 51″·9 | | cot. | = 9·9338347 |
| | $c$ | = 446·1423 feet | | log. | = 2·6494734 |
| | $c'$ | = 188·3482 | | | |
| | $c' + c$ | = 634·4905 | | | |
| | $t$ | = 634·4905 | | | |

In this example it appears that the tangent point sought is at T.

NOTE. The reader will perceive that this example is taken from the results of the example next but one preceding, and is intended as a test to both.

[FIG. 11.]

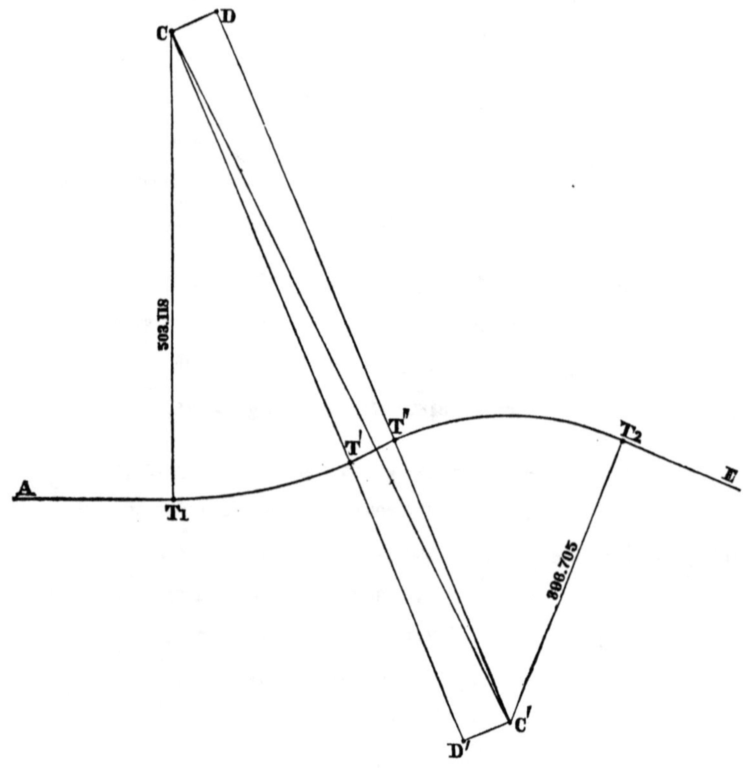

(**31**) Another form of reversed curve (if the expression be a proper one) for uniting tracks having different, or like bearings, is where you have the relative position of the tangent points from whence the curves commence in the given tracks, with the bearings of said tracks, or (which is the same thing) the tangent lines, and the radii of the curves given or required by the contour of the surface, or other considerations which may govern the location.

In order to give this problem a practical character, we copy from a case which actually occurred in the practice of the writer. We shall not give all the preliminary surveying which was deemed necessary to guide us in the location, (which had been in amount considerable,) but will only state that many lines were run and measured in various directions, to such points as we were desirous of knowing the relative situations of, and the traverses were worked up; or, in other words, the relative situations of these points were computed in northings and southings, eastings and westings. We copy from those tables such data as we shall find necessary to enable us to explain and solve the problem, and render our computations intelligible.

The position of first tangent point, 174·306 feet northing, 159·617 feet easting.

The bearing of tangent line from first tangent point, N. E. 85° 44′ 35″.

The bearing of radius from first tangent point, S. E. 4° 15′ 25″, and its length = 503·118 feet, log. = 2.7016699

The position of the centre of the curve formed by the above radius = 327·424 feet southing, 196·963 feet easting.

The position of second tangent point, 64·735 feet northing, 509·235 feet westing.

The bearing of tangent line from second tangent point, N. W. 61° 11′ 53″.

The bearing of radius from second tangent point, N. E. 28° 48′ 07″ and its length, 306·705 feet, log. = 2·5984678

The position of the centre of the curve formed by the above radius = 412·364 feet northing; 318·109 feet westing.

Representing the first tangent point by $T_1$, and the second tangent point by $T_2$, and the interior tangent points by T′ and T″, (see Fig. 11,) the centre of the curve with the radius of 503·118 feet by C, and the centre of the curve with the radius of 396·705 feet by C′; then will C T′ = 503·118 feet, and C′ T″ = 396·705 feet.

Having constructed our diagram in conformity with the data given, we commence by finding the distance C C′. We have given the

| | | |
|---|---|---|
| Position of C | = 327·424 feet southing, | and 196·963 feet easting, |
| " " C′ | = 412·364 feet northing, | " 318·109 feet westing, |
| Diff. of northing, | 739·788 feet | 515·072 feet diff. of westing. |

Having obtained the difference of northings and westings between C and C″, we have this analogy, to find the bearing from one to the other, viz. Assuming the distance C C′ = radius, and the difference in the northings as a cosine, and the difference of westings as a sine; then, representing these functions, viz., radius by R, the sine by $s$, the cosine by $c$, the bearing by B,

$$\text{We have } c : s :: \text{R} : \text{tan. B} = \frac{s}{c}$$

or, more practically to express the same thing, we have, tan. B = difference of westings, divided by the difference of northings, and

$$\text{Sin. B} : \text{westing} :: \text{R} : \text{C C}' = \frac{\text{westings}}{\text{sin. B}}$$

$$\text{or, Cos. B} : \text{northing} :: \text{R} : \text{C C}' = \frac{\text{northings}}{\text{cos. B}}$$

Having thus found C C′, we compare it with the sum of the radii, viz., C T′ + C′ T″; and if C T′ + C′ T″ be found greater than C C′, the assumed radii will be too great for the relative situation of things; but if C T′ + C′ T″ be found less than C C′ the two curves will not run into each other, and must be connected by a piece of straight line. The determination of this straight line is the object of the present investigation.*

For an example of computation we have given the

| | | | | |
|---|---|---|---|---|
| Position of C | = 327·424 feet southing, | and | 196·963 feet easting, | |
| and of C′ | = 412·364 " northing, | " | 318·109 " westing, | |
| Diff. of northings, | 739·788 " | " | 515·072 " diff. of westings. | |

| | | | |
|---|---|---|---|
| Difference of westings | = 515·072 | ........... log. | = 2·7118680 |
| " " northings | = 739·788 | ........... log. | = 2·8691072 |
| Bearing from C to C′ = B | = N. W. 34° 50′ 50″·24 | tan. | = 9·8427608 |

| | | | |
|---|---|---|---|
| B = 34° 50′ 59″·24 co. ar. sin. | = 0·2430665 | co. ar. cos. | = 0·0858273 |
| Westing = 515·072 log. | = 2·2118680 | northing 739·788 log. | = 2·8691072 |
| From C to C′ = 901·435 log. | = 2·9549345 | Proof, log. | = 2·9549345 |

The given radii are

| | |
|---|---|
| From $T_1$ | = 503·118 feet |
| " $T_2$ | = 396·705 |
| C T′ + C T″ = $r + r$ | = 899·823 |
| C C′ | = 901·435 |
| Difference | 1·612 feet |

C C′ being longer than C T′ + C′ T″, it is evident it will require several feet of straight line to connect them.

* It is obvious that if C C′ and (C T′ + C′ T″) are found the same length, the two will run into each other, and form perfect reverse curves.

Representing C T′ + C′ T″ by $r + r'$, and C C′ by H, we have

H : R :: $r + r'$ : sin. C or sin. C′ $= \frac{r + r'}{H}$

Sin. C : $r + r'$ :: cos. C · D C = D′ C′ = T′ T″ = cot. C $(r + r')$

| | | | | |
|---|---|---|---|---|
| ∴ $r + r'$ | = 899·823 | ............ | log. = | 2·9541571 |
| H | = 901·435 co. ar. | ............ | log. = | 7·0450655 |
| C | = 86° 34′ 22″ | ............ | sin. = | 9·9992226 |
| And C | = 86° 34′ 22″ | ............ | cot. = | 8·7772380 |
| $r + r$ | = 899·823 | ............ | log. = | 2·9541571 |
| T′ T″ | = 53·888 feet | ............ | log. = | 1·7313951 |

Having thus found the length of the straight line connecting the two curves, it becomes a matter of considerable interest to know the magnitude of the centre angle belonging to each curve respectively.

We found the bearing from C to C′ to be N. W. 34° 50′ 50″·24; and the angle C in the triangle C C′ D = 86° 34′ 22″, the complement to which will be 3° 25′ 38″.

| | | |
|---|---|---|
| Then, the bearing from C to C′ ..................... | = N. W. | 34° 50′ 50″·24 |
| It is obvious that if we subtract the complement D′ C C′ | = | 3° 25′ 38″ |
| Will leave the bearing C T′ and C′ T″......... | = N. W. | 31° 25′ 12″·24 |
| The bearing C $T_1$ being ......................... | = N. W. | 4° 15′ 25″ |
| Gives the centre angle $T_1$ C T′ .............. | = | 27° 09′ 47″·24 |
| Again, C′ T″ bearing (as above).................... | = N. W. | 31° 25′ 12″·24 |
| And the radius $T_2$ C bears........................ | = N. E. | 28° 48′ 07″ |
| Gives for the centre angle $T_2$ C′ T″ ........... | = | 60° 13′ 19″·24 |

Such further elements as may be deemed useful in the location might be readily obtained by such of the preceding formula as may be found applicable.

**(32)** There will doubtless arise in practice a great variety of cases, or conditions requiring reverse curves, many of them requiring formula entirely different from those we have been investigating, while there are many others which will require merely some slight modifications. But, to repeat what has been more than once stated, it is not our purpose to pursue these investigations until the subject is exhausted, but only to present those cases which we have presumed would most frequently occur in practice. We, however, have another class of curves, the greater portion of them reversing curves, viz., turnouts and side tracks, which may be worthy of consideration. We will therefore proceed to the investigation of formulæ for obtaining the necessary elements for locating them, and in the same connection will endeavor to ascertain the magnitude of the angles the rails make with each other at the points of crossing, or, in other words, the dimensions and form of the frogs necessary to be used to best suit each particular case.

**(33)** Before we proceed with the investigations, I would make a few remarks upon the switch-bar. The switching of the bar, preparatory to turning a train upon a side track, becomes an important element in our investigation. We have no doubt this element would be considered by persons who have not fully investigated the subject, as unnecessarily complicating our formulæ, and of course our computations.

The first consideration in preparing the switch, is to ascertain the smallest amount of sliding motion, that will answer to pass the wheels, and, at the same time, give firmness and security to the ends of the rails. The pattern of rails generally used in Massachusetts requires a movement of about five inches, and the pattern for the switch castings used to secure the ends of the rails, and to give firmness and stability to the structure, are nearly uniform in their dimensions; hence, whether the switch rail be long or short, whether the turnout be of large or small radius, the switching, or movement of the end of the bars, remains the same.

It may happen, however, that when the turnout curve is required to be exceedingly severe, and we desire to make the most of the room we have at command, that we determine by calculation the length the switch-bar (switching five inches, or the amount required by the castings) must be to make it exactly correspond to a portion of the intended curve, and the switch rails are accordingly cut to that length. But, if there is nothing to prevent the radius of the turnout from taking such length as may be deemed most desirable, it becomes the better policy to have the switch-bars as long as the bars with which the track is laid, or is being laid.

The longer the switch-bar is, the smaller will be the angle of deflection occasioned by switching; and the smaller the deflecting angle, the less the impediment to the passage of the engine and cars, and less springing of the bar than when the bar is shortened; and of course less liability to accident.

In general, the deflection of the switch-bar should not be greater than the deflection of the curve for the same length of arc. Cases will, however, occur, when the deflection of the switch-bars of the greatest length in use, will exceed the deflection of the like quantity of arc. These cases occur frequently at the connection of branches; and, in general, we may say (if this discussion be correctly based) that the switch-bar, when switched, should in all cases be considered the tangent line from whence the curve is to spring, or commence. It may, however, be neglected, when the switching exactly corresponds with the deflection of the same length of curve; but it will not in that case interfere with the accuracy of the calculations to then consider it as the tangent line. Cases may, however, occur, requiring the tangent lines to be continued beyond the end of the bar before the curve commences; but these cases will not often be met with.

(**34**) Having said thus much respecting switches, we commence our investigation by considering the most simple form of the turn-out, viz., from a straight track, with curves of equal radii.

NOTE. I would here state, for the information of the young engineer, that the side-track curves, when there is nothing to interfere, should be laid to a radius of, say from five to six hundred feet; but when the nature of things demand it, they may be laid to a radius as short as two hundred and fifty feet. If a radius still shorter is demanded, it becomes necessary to lay extra rails upon the inside of the curve, and as near the rails of the main track as they can be well secured, to assist in supporting the centre driving wheels of the engine, which would otherwise be sometimes unsupported, and would then cause the engine to run off the track. I hardly need to remark that the double rail will be useless when the engines have only one pair of driving wheels.

To proceed with the investigation. We first ascertain the relative position of the switch-bar, or the angle it makes with the main track.

[FIG. 12.]

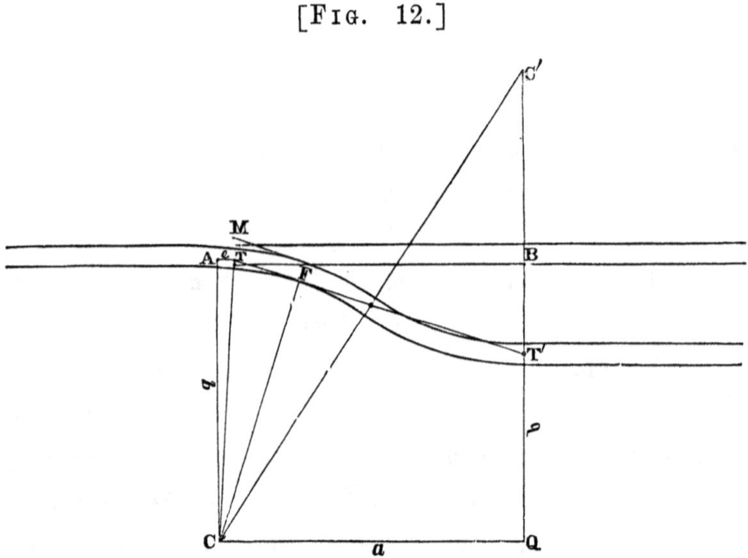

Let $S$ represent the length of the switch-rail, and $d$ the distance it slides; $Sw$ the switch angle, or the angle the switch-rail makes with the main track when it is switched. We then have

$$S : \mathrm{R} :: d \,.\, \sin. Sw = \frac{d}{S} \text{ or } \tan. Sw^{*} \qquad (32)$$

Having thus obtained the switch angle, we will now put $r =$ radius of the turnout; $a = \mathrm{C\,Q}$; $b = \mathrm{Q\,T'}$; $q = \mathrm{C\,A}$; $\delta =$ the distance between the track centres; $e = \mathrm{A\,T}$; $g = \mathrm{T\,B}$.

We have in the triangle A C T, to find $q$ and $e$.

$$\left.\begin{array}{l} \mathrm{R} : r :: \cos. Sw : q = r \,.\, \cos. Sw \\ \mathrm{R} : r :: \sin. Sw : e = r \,.\, \sin. Sw \end{array}\right\} \qquad (33)$$

Then will $b = q - \delta + d$, and

$$2\,r : \mathrm{R} :: (r + b) : \cos. \mathrm{C'} = \frac{r + b}{2r} \qquad (34)$$

$$\mathrm{Cos.\,C'} : (r + b) :: \sin. \mathrm{C'} : a = \tan. \mathrm{C'}\ (r + b) \qquad (35)$$

We also have $a - e = g$; and the angle $\mathrm{C} = \mathrm{C'} - Sw$.

We have now found the principal elements necessary for locating and marking the centre line of the turnout; whatever practice requires to fill up the details may readily be supplied from formulæ given in the preceding pages.

**(35)** The frog angle next claims our attention. Representing the distance between the rails, or in other words, the guage of the track, by $h$, we have $\mathrm{C\,F} = r + \frac{1}{2}\,h$; and $\mathrm{C} \oplus = r - \frac{1}{2}\,h + d$;

* To be strictly exact, we make use of the following analogy:

$$S : \mathrm{R} :: \tfrac{1}{2}\,d : \sin. \tfrac{1}{2}\,Sw = \frac{\frac{1}{2}\,d}{S}$$

but this formula is rather a refinement than otherwise, as either of the two first expressions are sufficiently exact for practice, and more convenient.

and the angle C ⊕ F = 90° + *Sw*. We now have, in the triangle C ⊕ F, $r + \frac{1}{2} h$ : sin. (90° + *Sw*) :: $r - \frac{1}{2} h + d$ : sin. F =

$$\frac{\sin. (90^\circ + Sw)\, (r - \frac{1}{2} h + d)}{(r + \frac{1}{2} h)} \qquad (36)$$

Then, drawing at F the tangent line F M, which of course must be at right angles to C F, it will be apparent that the frog angle M F ⊕ will be a complement angle to F as found above; wherefore, we have 90° — F = M F ⊕; and the angle at the centre C, will be equal to 180° — (⊕ + F;) or, which amounts to the same, C = M F ⊕ — *Sw;* and the chord, which we represent by *ch*, from the mouth of the switch upon the outside rail of the turnout track, to the point where the frog angle should be placed in the main track, may be ascertained by the following analogy:

$$\text{Sin. } \tfrac{1}{2}\,(180^\circ - C) : r + \tfrac{1}{2} h :: \sin. C : ch = \frac{\sin. C . (r + \frac{1}{2} h)}{\sin. \frac{1}{2}\,(180^\circ - C)} \qquad (37)$$

The chord just found will be of great convenience to the track-layers, as it will show them the proper place for the frog, which should be put into the main track when they are laying it down.

Having thus obtained our formulæ, we now proceed with an example of computation.

We will assume $r = 499{\cdot}725$ feet, which gives a deflection of 1° 26′ for a 25 ft. chord; $h = 4{\cdot}7$ feet; $d = 5$ inches; $S = 21$ feet; and $\delta = 11$ feet.

By formula (32) we have

| | | | |
|---|---|---|---|
| $d$ = 5 inches | ............ | log. = | 0·6989700 |
| $S$ = 252 inches co. ar. | ............ | log. = | 7·5985995 |
| $Sw$ = 1° 08′ 12″ | ............ | tan. = | 8·2975695 |

| | | | | |
|---|---|---|---|---|
| | $Sw = 1° 08' 12''$ | ............ | cos. = 9·9999145 | (33) |
| | $r = 499·725$ | ............ | log. = 2·6987307 | |
| | $q = 499·6262$ feet | ............ | log. = 2·6986452 | |
| Again, | $Sw = 1° 08' 12''$ | ............ | sin. = 8·2974820 | |
| | $r =$ | ............ | log. = 2·6987307 | |
| | $e = 9·9131$ | ............ | log. = 0·9962127 | |

| | | | |
|---|---|---|---|
| $2r = 999·449$ co. ar. | .......... | log. = 7·0002393 | |
| $r + b = 988·768$ | .......... | log. = 2·9950944 | |
| $C' = 8° 23' 02''·7$ | .......... | cos. = 9·9953337 | (34) |
| $C' = 8° 23' 02''·7$ | .......... | tan. = 9·1684391 | |
| $r + b =$ | .......... | log. = 2·9950944 | |
| $a = 145·7248$ feet | .......... | log. = 2·1635335 | (35) |
| $e = 9·9131$ | | | |
| $(a - e) = 135·8117 =$ | | | |

the distance on the main track from the mouth of the switch to a point opposite T′, T′ being off at right angles from the point.

$$C' = 8° 23' 02''·7$$
$$Sw = 1° 08' 12''$$
$$C' - Sw = C = 7° 14' 50''·7$$

| | | | | |
|---|---|---|---|---|
| $r + \frac{1}{2}h$ | $= 502·0746$ feet co. ar. | log. = 7·2992318 | |
| $90° + Sw$ | $= 91° 08' 12''$ | sin. = 9·9999145 | |
| $r + d - \frac{1}{2}h$ | $= 497·7912$ feet | log. = 2·6970473 | |
| F | $= 82° 25' 31''$ | sin. = 9·9961936 | (36) |

$$90° - F = \text{frog angle} = 7° 34' 29''$$
$$Sw = 1° 08' 12''$$
$$\text{Frog angle} - Sw = C'' = 6° 26' 17''; \; C'' =$$

the angle at C in the triangle T C F.

| | | | |
|---|---|---|---|
| $\frac{1}{2}(180° - C)$ | $= 86° 46' 51''.5$ co. ar. | sin. | $= 0.0006858$ |
| $r + \frac{1}{2}h$ | $= 502.0746$ feet | log. | $= 2.7007682$ |
| $C''$ | $= 6° 26' 17''$ | sin. | $= 9.0497178$ |
| $ch$ | $= 56.386$ feet | log. | $= 1.7511718 =$ |

chord distance from mouth of switch to the angle of the frog upon the outside rail of the turnout.

Recapitulation of the elements obtained, viz.,

Centre angles $C' = \ldots\ldots$ $8° 23' 02''.7$

" " $C = \ldots\ldots$ $7° 14' 50''.7$

Frog angle M F ⊕ $= \ldots\ldots$ $7° 34' 29''.0$

Chord distance from the mouth of switch to mouth of frog, (outside rail,) 56·386 feet.

**(36)** We find wanting the relative position of the point where the curves reverse. The formula will be

$$\text{Sin. } \tfrac{1}{2}(180° - C) : r :: \text{sin. } C : c = \frac{r \,.\, \text{sin. } C}{\text{sin. } \frac{1}{2}(180° - C)} \qquad (38)$$

wherein $c =$ the chord distance from the centre point between the mouth of the switch-bars when switched, and the point where the curve reverses.

Then, to find the chord of the reverse curve $= c'$, we have

$$\text{Sin. } \tfrac{1}{2}(180° - C') : r :: \text{sin. } C' : c' = \frac{\text{sin. } \frac{1}{2}(180° - C')}{r \text{ sin. } C'} \qquad (39)$$

representing by $c'$ in the foregoing analogy, the chord from the reversing point to the tangent point T′.

EXAMPLE OF COMPUTATION.

| | | | | |
|---|---|---|---|---|
| $\frac{1}{2}(180° - C)$ | $= 86° 22' 34''.65$ co. ar. | sin. | $= 0.0008692$ | |
| $r$ | $= 499.725$ feet | log. | $= 2.6987307$ | |
| $C$ | $= 7° 14' 50''$ | sin. | $= 9.1008914$ | |
| $c$ | $= 63.167$ feet | log. | $= 1.8004913$ | (38) |

| | | | |
|---|---|---|---|
| $\frac{1}{2}$ (180° — C′) = 85° 48′ 28″.65 co. ar. | sin. = 0·0011635 | | |
| $r$ = 499·725 feet | log. = 2·6987307 | | |
| C′ = 8° 23′ 02″·7 | sin. = 9·1631819 | | |
| $c'$ = 72·958 feet | log. = 1·8630761 | (39) | |

**(37)** To lay down the chords $c$ and $c'$, we commence by placing the theodolite at the point in the centre of the mouth of the switch, when switched; then, pointing the telescope in the direction towards B, in a range parallel with the main track, lay off an angle towards the side of the road to which the switch-bar switches = (90° — $Sw$) + $\frac{1}{2}$ (180° — C,) then measure from the instrument the distance $c$ for the reversing point. Then, moving the theodolite to the reversing point, and directing the telescope to the point just left, (to wit, in the centre of the mouth of the switch,) lay off an angle towards the main track = 180° — $\frac{1}{2}$ (180° — C) + $\frac{1}{2}$ (180 °— C′,) and measure the distance $c'$ for the tangent point of the turnout. The remainder of the laying out may be performed by deflections, or other methods, as explained in the foregoing pages.

We have now, I think, obtained every element necessary for locating and marking out a turnout from a straight track, and for making a frog pattern to suit.

N. B. If the tangent points, the reversing point, and the place for the frog be distinctly and properly marked, and the rails properly curved, a skilful tracklayer would put in a turnout without further laying out.

**(38)** We now proceed to the investigation of formulæ for determining the radius of a turnout from a straight track suited to a given frog.

[Fig. 13.]

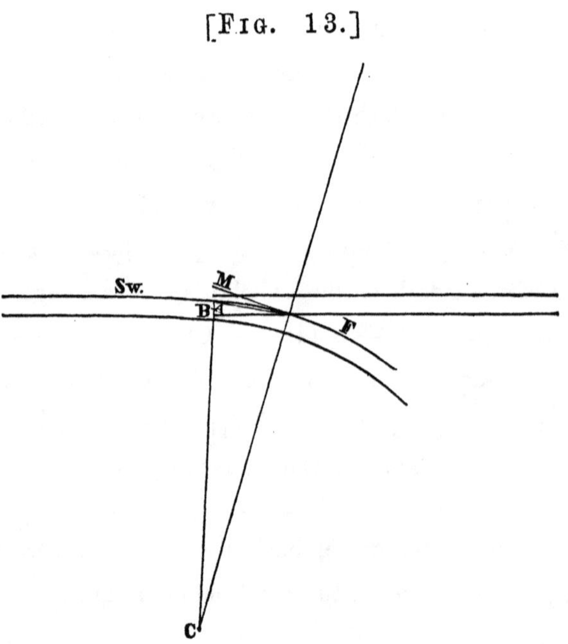

Let *Sw* represent the switch angle.

*Fr* " " frog "

C " " centre "

A " one of the equal angles in isosceles triangle A F C.

$h$ " the distance between the rails.

$d$ " " " the switch slides.

$r$ " " radius.

We now have in the triangle B C F, (See Fig. 13,) the angle at B $= 90° + Sw$; the angle at F $= 90° - Fr$; and the angle at C $= 180° - (B + F.)$ Then, in the triangle A B F, representing F by $F_2$, and B by $B_2$, we have the angle $A = \frac{1}{2}(180° - C;)$ and the angle $F_2 = (A - F,)$ F being $= (90° - Fr)$ as above, and the angle $B_2 = 90° - Sw$.

Having thus determined the angles, we have in the triangle F A B, $AB = \frac{h-d}{\sin. B_2}$; that is, $\sin. B_2 : h - d :: R : AB$; then, representing F A by $ch$, and A B by $w$, we have

$$\text{Sin. } F_2 : w :: \sin. B_2 : ch = \frac{w \sin. B_2}{\sin. F_2} = \frac{h - d \,.\, \sin B_2}{\sin. B_2 \,.\, \sin. F_2} = \frac{h-d}{\sin. F_2}$$

$$\text{Sin. } C : ch :: \sin. A : r + \tfrac{1}{2} h = \frac{ch \sin. A}{\sin. C} = \frac{(h - d) \,.\, \sin. A}{\sin. F_2 \,.\, \sin. C}$$

then, by subtracting $\frac{1}{2} h$ we have $r$.

### EXAMPLE OF COMPUTATION.

Let $Fr = 7° 34' 29''$; $Sw = 1° 08' 12''$; $h = 4·7$ feet; $d = 0·4166$; then,

| | | |
|---|---|---|
| 90° | 90° | 180° |
| $Fr = 7° 34' 29''$ | $Sw = 1° 08' 12''$ | $C = 6° 26' 17''$ |
| $90° - Fr = 82° 25' 31'' = F$ | $90° + Sw = 91° 08' 12'' = B$ | 2)173° 33′ 43″ |
| $91° 08' 12'' = B$ | 90° | $A = 86° 46' 51''·5$ |
| $180° - (B + F) = 6° 26' 17'' = C$ | $Sw = 1° 08' 12''$ | $F = 82° 25' 31''·$ |
| 180° 00′ 00″ | $90° - Sw = 88° 51' 48'' = B_2$ | $F_2 = 4° 21' 20''·5$ |

[Fig. 14.]

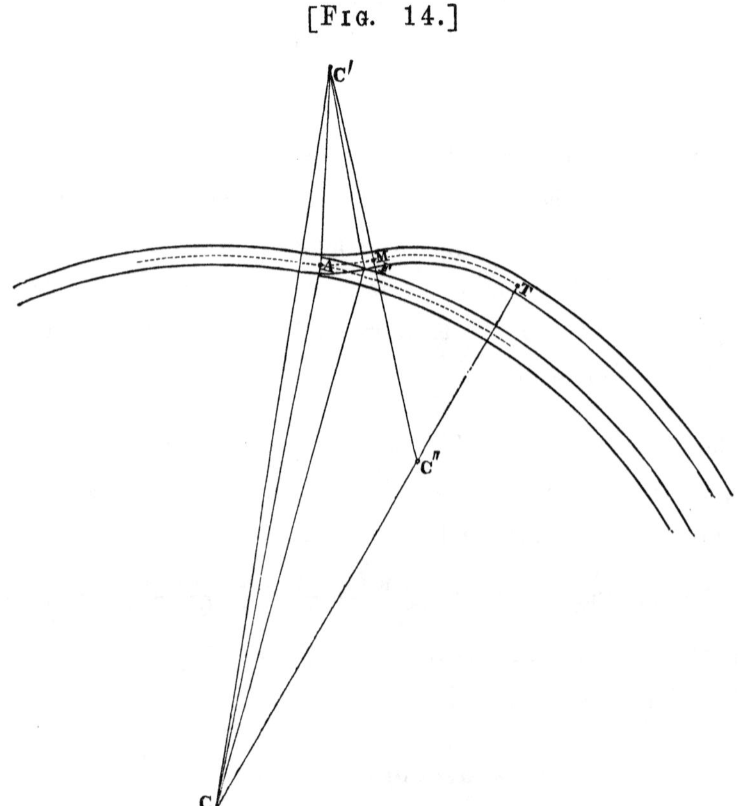

| | | | |
|---|---|---|---|
| $F_2$ | $= 4^\circ\ 21'\ 20''.5$ co. ar. | ........ sin. | $= 1.1194836$ |
| C | $= 6^\circ\ 26'\ 17''$ co. ar. | ........ sin. | $= 0.9502822$ |
| A | $= 86^\circ\ 46'\ 51''.5$ | ........ sin. | $= 9.9993142$ |
| $h - d$ | $= 4.2834$ | ........ log. | $= 0.6317886$ |
| $r + \frac{1}{2}h$ | $= 502.19$ feet | ........ log. | $= 2.7008686$ |
| $\frac{1}{2}h$ | $= 2.35$ | | |
| $r$ | $= 499.84$ | | |

We have thus found the radius $= 499.84$; it was intended as a reverse of the previous problem, which gave 499·725 feet. The difference, it will be perceived, is a mere trifle, and is owing to the loss of small fractions in the frog angle, and by using tables of limited extent.

**(39)** The next form of a turnout which we shall consider, is one which shall turn out upon the outside of a curve.

Retaining the same length of switch we had in our preceding calculation, of course the switch angle will remain the same. Then, representing the switch angle by *Sw*; the slide motion by $d$; the radius of the main track by $r$; the radius of the turnout by $r'$; the radius of the side track by $r''$; and the line C C′ by $a$, we have in the small triangle C A C′, two sides, and an included angle, viz., the angle at A $=$ the supplement of the switch angle *Sw*, and C′ A $= r'$; C A $= r$, to find the remaining side $a$, and the angles C and C′.

This problem has been so often investigated, and is so well understood, that I have deemed it unnecessary to give an investigation. We however give a formula in connection with the investigation of the turnouts, so that the computer is enabled, without being obliged to look up at the time, elementary works to supply the deficiency of such papers, or to refresh his memory, where, perchance, he may be somewhat in doubt.

In the following formula we shall use the symbols by which we represent the triangle under consideration.

$$\text{Tan. } X = \frac{\text{tan. } \frac{1}{2} (180^\circ - A) \, [(r+d) \backsim r']}{r' + (r+d)}; \text{ and } X + \tfrac{1}{2} (180^\circ - A) = C';$$

$$\text{and } X - \tfrac{1}{2} (180^\circ - A) = C.$$

$$\text{And sin. } C' : r + d :: \text{sin. } A : a = \frac{(r+d) \,.\, \text{sin. } A}{\text{sin. } C'} \qquad (40)$$

or, by way of proof, we have

$$\text{Sin. } C : r' :: \text{sin. } A : a = \frac{\text{sin. } A \,.\, r'}{\text{sin. } C}$$

which, if our previous computations have been correctly performed, the results of this and the preceding analogy will be alike.

**(40)** Having solved the small triangle, we next endeavor to find the magnitude of the frog angle, and its relative position in the main track. For the accomplishment of which object, we have in the triangle C F C′ the three sides, to obtain their angles.

Having obtained their angles, we then, by the solution of the triangle C A F, obtain the chord, which we shall represent by *ch*, and which will give us the distance of the frog angle in the main track from the mouth of the switch upon the outside rail of the turnout, or from A to F. We also give the following formula, without going into a general investigation, using the symbols by which we represent the lines of the present triangle under consideration. Representing the guage of the track by $h$, we have, in the triangle C F C′, the line C C′ $= a$, obtained by (40;) the line C F $= r + \frac{1}{2} h$, which we represent by $b$; the line C′ F $= r' + \frac{1}{2} h$, which we represent by $c$.

If we now put $p = \frac{1}{2}(a + b + c)$ we have

$$\text{Tan. } \tfrac{1}{2}\text{ F} = \left(\frac{(p-b).(p-c)}{p\,(p-a)}\right)^{\frac{1}{2}}; \text{ and tan. } \tfrac{1}{2}\text{ C}' = \left(\frac{(p-a).(p-c)}{p\,(p-b)}\right)^{\frac{1}{2}};$$

$$\text{and tan. } \tfrac{1}{2}\text{ C} = \left(\frac{(p-a).(p-b)}{p.(p-c}\right)^{\frac{1}{2}}; \qquad (41)$$

and $180° - \text{F} = \text{frog angle} = \text{F}'$.

Substituting $\text{C}'_2$ for $\text{C}'$ as obtained by formula (40,) we have

$$\text{Sin. } \tfrac{1}{2}\,[180° - (\text{C}' - \text{C}'_2)] : c :: \text{sin. } (\text{C}' - \text{C}'_2) : ch = \frac{c.\text{ sin. } (\text{C}' - \text{C}'_2)}{\text{sin. } \frac{1}{2}\,(180° - \text{C}' - \text{C}'_2)}$$

or, probably, in practice, the following may be substituted with convenience, viz.,

$$\text{R} : 2c :: \text{sin. } \tfrac{1}{2}\,(\text{C}' - \text{C}'_2) : ch = \text{sin. } \tfrac{1}{2}\,(\text{C}' - \text{C}'_2)\; 2c \qquad (42)$$

**(41)** The next step in our investigation will be to ascertain the reversing point, M; and the terminus of the reverse curve, T. For this purpose, in the triangle C C′ C″ we have three sides, viz., the side $\text{C C}' = a$, from (40;) side $\text{C}'\text{ C}'' = 2\,r'$; and the side $\text{C C}'' = \text{C T} - r'$; and $\text{C T} = r + \delta$; therefore, $\text{C C}'' = r + \delta - r'$.

Representing by $\delta$ the distance between centre lines of the main and side tracks, and substituting $b$ for C C″, and $c$ for C′ C″, we obtain the angles by (41;) and then, to find A M, which we represent by $ch'$, we have

$$\text{R} : 2r' :: \text{sin. } \tfrac{1}{2}\,(\text{C}' - \text{C}'_2) : ch' = 2r'.\text{ sin. } \tfrac{1}{2}\,(\text{C}' - \text{C}'_2) \qquad (43)$$

Then, to find the chord M T, which we represent by $ch''$, we have

$$\text{R} : 2r' :: \text{sin. } \tfrac{1}{2}\,(180° - \text{C}'') : ch'' = 2r'.\text{ sin. } \tfrac{1}{2}\,(180° - \text{C}'') \qquad (43)$$

To lay off these chords, we place the instrument at the centre of the mouth of the switch, when switched, pointing in the direction parallel to the tangent of the curve of the main track, (the tangent to the curve of the turnout will be found to vary from the tangent of the main track in amount equal to the switch angle;) lay off an angle towards the turnout side of the road $= 90° - Sw + \frac{1}{2}(180° - C';)$ then, measure the chord *ch* to M, the reversing point; then, moving the instrument to M, and pointing it to the station between the mouth end of the switch-bars just left, lay off an angle on the side towards the main track equal to $180° - \frac{1}{2}(180° - C') + \frac{1}{2}(180° - C'';)$ then measuring the chord *ch''* to the tangent point T.

We think that further details of the method of locating the curves need not be here given, the principles having been fully explained in the foregoing pages.

To proceed with an example of computation. We put $r =$ 5729·597 feet; $r' = 499·725$ feet; $h = 4·7$ feet; $d = 0·416$ feet; $\delta = 11$ feet; $Sw = 1° \ 08' \ 12''$; and of course A =

$$\begin{array}{r} 178° \ 51' \ 48''. \\ \hline 2) \ 1° \ 08' \ 12'' \\ \hline 0° \ 34' \ 06'' \end{array} \qquad (40)$$

| | | | |
|---|---|---|---|
| $r + d$ | = 5730·013 | | |
| $r'$ | = 499·725 | | |
| $(r + d) + r'$ | = 6229·738 | co. ar. | log. 6·2055302 |
| $(r + d) \backsim r'$ | = 5230·288 | | log. 3·7185256 |
| $\frac{1}{2}(180° - A)$ | = 0° 34' 06'' | | tan. 7·9964947 |
| X | = 0° 28' 37''·772 | | tan. 7·9205505 |
| C' | = 1° 02' 43''·772 | | |
| C | = 05' 28''·228 | | |
| A | = 178° 51' 48'' | | |
| Proof | 180° 00' 00''·000 | | |

| | | | | | |
|---|---|---|---|---|---|
| C′ | $= 1^\circ\ 02'\ 43''\cdot772$ co. ar. sin. | $= 1\cdot7388259$ | $C = 0^\circ\ 05'\ 28''\cdot228$ co. ar. sin. | $= 2\cdot7982497$ |
| $r + d$ | $= 5730\cdot013$ feet log. | $= 3\cdot7581556$ | $499\cdot725$ feet log. | $= 2\cdot6987311$ |
| A | $= 178^\circ\ 51'\ 48''$ sin. | $= 8\cdot2974820$ | | $8\cdot2974820$ |
| $a$ | $= 6229\cdot647$ feet log. | $= 3\cdot7944635$ | | $3\cdot7944628$ |

NOTE. By a more strict computation, the second analogy gave the same results as the first.

$$r = 5729\cdot597 \qquad r' = 499\cdot725$$
$$\tfrac{1}{2}h = 2\cdot350 \qquad \tfrac{1}{2}h = 2\cdot350$$
$$r + \tfrac{1}{2}h = 5731\cdot947 = b \qquad r' + \tfrac{1}{2}h = 502\cdot075 = c \qquad (41)$$

$a = 6229\cdot641$
$b = 5731\cdot947$
$c = 502\cdot075$
$2)\,12463\cdot663$

| | | |
|---|---|---|
| $p$ | $= 6231\cdot831$ | log. $= 3\cdot7946157$ |
| $p' - a$ | $= 2\cdot190$ | log. $= 0\cdot3404441$ |
| $p - b$ | $= 499\cdot884$ | log. $= 2\cdot6988693$ |
| $p - c$ | $= 5729\cdot756$ | log. $= 3\cdot7581362$ |

| | |
|---|---|
| $p$ co. ar. log. | $= 6\cdot2053843$ |
| $p - a$ co. ar. log. | $= 9\cdot6595559$ |
| $p - b$ log. | $= 2\cdot6988693$ |
| $p - c$ log. | $= 3\cdot7581362$ |
| | $2)\,22\cdot3219457$ |
| $\tfrac{1}{2}F = 86^\circ\ 03'\ 04''\cdot44$ tan. | $= 11\cdot1609728$ |
| 2 | |
| $F = 172^\circ\ 06'\ 08''\cdot88$ | |
| $7^\circ\ 53'\ 51''\cdot12$ = the frog angle. | |

| | |
|---|---|
| $p$ co. ar. log. | $= 6\cdot2053843$ |
| $p - b$ co. ar. log. | $= 7\cdot3011307$ |
| $p - c$ log. | $= 3\cdot7581362$ |
| $p - a$ log. | $= 0\cdot3404441$ |
| | $2)\,17\cdot6050953$ |
| $\tfrac{1}{2}C' = 3^\circ\ 37'\ 53''\cdot46$ tan. | $= 8\cdot8025476$ |
| 2 | |
| $C' = 7^\circ\ 15'\ 46''\cdot92$ | |

| | |
|---|---|
| $p$ co. ar. log. | $= 6\cdot2053843$ |
| $p - c$ co. ar. log. | $= 6\cdot2418638$ |
| $p - a$ log. | $= 0\cdot3404441$ |
| $p - b$ log. | $= 2\cdot6988693$ |
| | $2)\,15\cdot4865615$ |
| $\tfrac{1}{2}C = 0^\circ\ 19'\ 02''\cdot09$ tan. | $= 7\cdot7432807$ |
| 2 | |
| $C = 0^\circ\ 38'\ 04''\cdot18$ | |

| | | |
|---|---|---|
| F = | $172^\circ\ 06'\ 08''\cdot88$ | Proof by adding angles. |
| C′ = | $7^\circ\ 15'\ 46''\cdot93$ | |
| C = | $0^\circ\ 38'\ 04''\cdot19$ | |
| | $180^\circ\ 00'\ 00''\cdot00$ | |

$$
\begin{array}{lll}
C' & = 7^\circ\ 15'\ 46''\cdot 93 & \\
C'_2 & = 1^\circ\ 02'\ 43''\cdot 77 & \\
 & 2)\overline{6^\circ\ 13'\ 03''\cdot 16} & \\
\tfrac{1}{2}(C' - C'_2) & = 3^\circ\ 06'\ 31''\cdot 58 & \sin. = 8\cdot 7342531 \\
r + \tfrac{1}{2} h & = 502\cdot 075 \text{ feet} & \log. = 2\cdot 7007686 \\
 & \quad 2\cdot & \log. = 0\cdot 3010300 \\
ch & = 54\cdot 457 \text{ feet} & \log. = 1\cdot 7360517
\end{array}
\qquad (42)
$$

$$
\begin{array}{lll}
a & = 6229\cdot 641 & \\
r + \delta - r' = b & = 5240\cdot 872 & \\
2r = c & = 999\cdot 450 & \\
 & 2)\overline{12469\cdot 963} & \\
p & = 6234\cdot 981 & \log. = 3\cdot 7948352 \\
p - a & = 5\cdot 340 & \log. = 0\cdot 7275413 \\
p - b & = 994\cdot 109 & \log. = 2\cdot 9974341 \\
p - c & = 5235\cdot 531 & \log. = 3\cdot 7189608
\end{array}
$$

$$
\begin{array}{ll}
p & \text{co. ar. log.} = 6\cdot 2051648 \\
p - a & \text{co. ar. log.} = 9\cdot 2724587 \\
p - b & \text{log.} = 2\cdot 9974341 \\
p - c & \text{log.} = 3\cdot 7189608 \\
 & 2)\overline{22\cdot 1940184} \\
85^\circ\ 25'\ 37''\cdot 62 & \tan. = 11\cdot 0970092 \\
 & 2 \\
C'' = 170^\circ\ 51'\ 15''\cdot 24 &
\end{array}
$$

$$
\begin{array}{ll}
p & \text{co. ar. log.} = 6\cdot 2051648 \\
p - b & \text{co. ar. log.} = 7\cdot 0025659 \\
p - c & \text{log.} = 3\cdot 7189608 \\
p - a & \text{log.} = 0\cdot 7275413 \\
 & 2)\overline{17\cdot 6542328} \\
3^\circ\ 50'\ 32''\cdot 15 & \tan. = 8\cdot 8271164 \\
 & 2 \\
C' = 7^\circ\ 41'\ 04''\cdot 30 &
\end{array}
$$

$$
\begin{array}{ll}
p & \text{co. ar. log.} = 6\cdot 2051648 \\
p - c & \text{co. ar. log.} = 6\cdot 2810392 \\
p - a & = 0\cdot 7275413 \\
p - b & = 2\cdot 9974341 \\
 & 2)\overline{16\cdot 2111794} \\
0^\circ\ 43'\ 50''\cdot 21 & \tan. = 8\cdot 1055897 \\
 & 2 \\
C' = 1^\circ\ 27'\ 40''\cdot 42 &
\end{array}
$$

$$
\begin{array}{ll}
C'' = & 170^\circ\ 51'\ 15''\cdot 25 \\
C' = & 7^\circ\ 41'\ 04''\cdot 31 \\
C = & 1^\circ\ 27'\ 40''\cdot 43 \\
 & \overline{180^\circ\ 00'\ 00''\cdot 00}
\end{array}
$$

$$
\begin{array}{lll}
\tfrac{1}{2}(180^\circ - C'') & = 4^\circ\ 34'\ 22''\cdot 37 & \sin.\ 8\cdot 9016060 \\
r & = 499\cdot 725 \text{ feet} & \log.\ 2\cdot 6987311 \\
 & \quad 2\cdot 000 & \log.\ 0\cdot 3010300 \\
ch'' & = 79\cdot 683 \text{ feet} & \log.\ 1\cdot 9013671
\end{array}
\qquad (43)
$$

We have thus completed the computations necessary to find all the leading or principal elements of a turnout upon the outside of a curve in the main track; and whatever of unexplained detail may be required can be computed in the field, as the computations will be both short and simple.

Let us now reverse the problem by supposing that we possess a frog of given dimensions, and are desirous to make it serve us in a turnout from the outside of a curve in the main track, whose radius of course we know. It will then become necessary to ascertain a radius for the turnout which will be suited to, or compare with, the angle of the frog.

(**42**) Without further remark we will proceed to the investigation of a formula. To render this investigation plain to the understanding, it may be necessary to become rather more particular in describing the figure or diagram upon which it is based, than it has heretofore been our custom. Making use of the same notation of the preceding problem, as far as applicable, we will commence the construction of the figure at the point where the angle of the frog is to be placed in the outside rail of the main track, viz., at F; from thence we draw a line to C $= r + \frac{1}{2} h$, for the radius of the outside rail of the curve in the main track, and describe a portion of the arc; then, from the same centre, with a radius $= r + \frac{1}{2} h + d$, describe the arc $S s$; and with a radius $= r - \frac{1}{2} h$, describe the inner rail of the main curve. At F lay off an angle from F C = the frog angle + the switch angle, and in accordance therewith draw the line F C″ $= r - \frac{1}{2} h + d$; and then, from the point C″ as a centre, with a radius $= r + \frac{1}{2} h$, describe the arc A S; the intersection of the arc with the little arc $S s$ at S will be the place

[Fig. 15.]

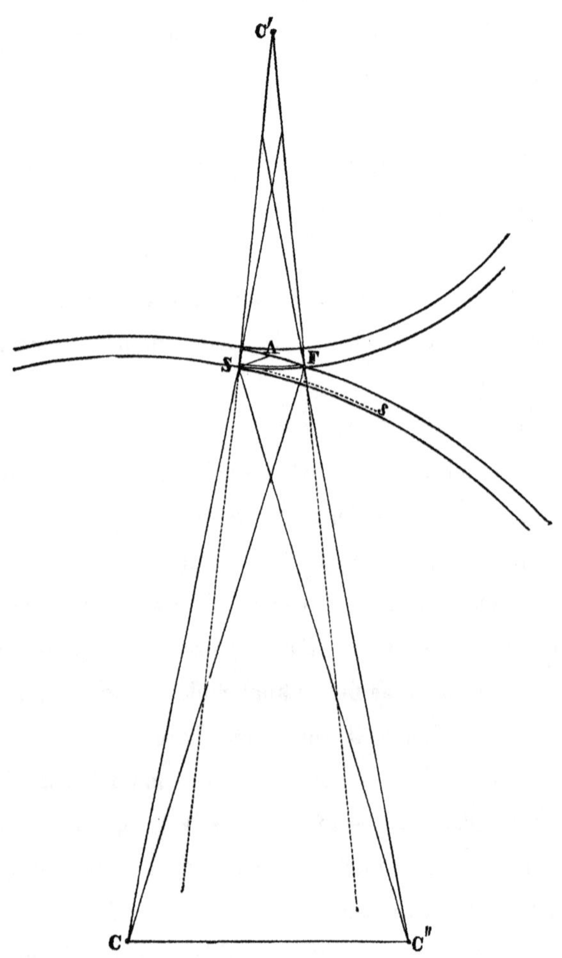

of the mouth end of the switch-rails, when switched; then, draw the radius from C to S, and continue the same indefinitely on the opposite side; from S draw the line S C′, making an angle with the continued radius = the switch angle; then, continue the line C″ F indefinitely, and draw the line F C′, making an angle with the continued line F C″ = the switch angle; the intersection of the lines S C′ with F C′ at C′ will determine the length of the radius of the turnout. If we now unite S C″ and C C″, we shall have a symmetrical figure containing two triangles, S C C″ and F C″ C, which are similar and equal. We shall have also the triangles S C F and F C″ S, which are similar and equal.

**(43)** Having thus completed our figure, we commence our investigation by endeavoring first to find the angle S C F.

We have the angle C S C″ = the angle C F C″, and the angle C F C″ as before stated = the frog angle + the switch angle. In the triangle C S C″ we have the angle S = the frog + the switch angle; and the side S C $= r - \frac{1}{2} h + d$; the side S C″ $= r + \frac{1}{2} h$. As we have before stated, the angles S C C″ and F C″ C are equal; hence, it is obvious that the angle sought, viz., S C F, is = to the difference between the angles S C″ C and S C C″; to find which, we have $(r + \frac{1}{2} h) + (r - \frac{1}{2} h + d) : (r + \frac{1}{2} h) \backsim (r - \frac{1}{2} h + d)$ :: tan. $\frac{1}{2}$ [180° — $(Fr + Sw)$] : tan. X, and 2 X = C, the angle sought. (44)

Having found the angle C, our direct course would be to find the angles S and F, and the side S F in the triangle C S F, which could readily be done by formulæ similar to the above, viz., (44;) but, believing the following to be more convenient, we pass that by.

N

We therefore have, in the quadrilateral figure C S C′ F, the angle at C = 2 X; the angle at S = 180° + the switch angle; the angle at F = (180° — frog angle,) and the angle at C′ = the explementary angle, or which shall make the sum of all the angles = 360°. It is now apparent that it will be convenient to represent by the letters C S C′ F, the angles belonging to three distinct figures, viz., the angles of the quadrilateral just named; the angles of the triangle C S F, and of the triangle C′ S F. For the purpose of preventing confusion, when we use the letters to denote an angle of the quadrilateral, they will not be accompanied by any distinguishing mark. When to denote an angle in the triangle C S F, they will be marked thus, $C_1$ $S_1$ $F_1$; and when to denote an angle in the triangle C′ S F, they will be marked thus, $C'_2$ $S_2$ $F_2$. We shall also denote the frog angle by *Fr*, and the switch angle by *Sw*.

To proceed with the investigation, we have

$$\tfrac{1}{2}\,(180° - C'_2) = S_2 = F_2\text{; and } F - F_2 = F_1\text{;}$$

$$\text{and } S - S_2 = S_1.$$

Having thus obtained all the angles of both triangles, we have

$$\text{Sin. } S_1 : r + \tfrac{1}{2}h :: \text{sin. } C_1 : S\,F = \frac{(r + \frac{1}{2}h)\,.\,\text{sin. } C_1}{\text{sin. } S_1} \qquad (45)$$

Substituting *c* for S F, we have an equal expression, which we frequently use by way of proof to our work, viz.,

$$\text{Sin. } F_1 : r - \tfrac{1}{2}h + d :: \text{sin. } C_1 : c = \frac{(r - \frac{1}{2}h + d)\,.\,\text{sin. } C_1}{\text{sin. } F_1} \qquad (45)$$

We next have sin. $C'_2$ : $c$ :: sin. $F_2$ or $S_2$ : $r'$ = the radius of the turnout sought. (46)

### EXAMPLE OF COMPUTATION.

We will suppose $r = 5729 \cdot 597$ feet; $Fr = 7° \, 53' \, 51'' \cdot 12$; $Sw = 1° \, 08' \, 12''$, to find the radius of turnout $r'$.

| | |
|---|---|
| $Fr$ | $= 7° \, 53' \, 51'' \cdot 12$ |
| $Sw$ | $= 1° \, 08' \, 12''$ |
| $Fr + Sw$ | $= 9° \, 02' \, 03'' \cdot 12$ |
| | $2)170° \, 57' \, 56'' \cdot 88$ |
| $\frac{1}{2}(180° - [Fr + Sw])$ | $= 85° \, 28' \, 58'' \cdot 44$ |

| | | | |
|---|---|---|---|
| $r + \frac{1}{2}h$ | $5731 \cdot 947$ | | |
| $r - \frac{1}{2}h + d$ | $5727 \cdot 663$ | | |
| $(r + \frac{1}{2}h) + (r - \frac{1}{2}h + d)$ | $= 11459 \cdot 610$ | co. ar. log. | $= 5 \cdot 9408302$ |
| $(r + \frac{1}{2}h) \backsim (r - \frac{1}{2}h + d)$ | $= 4 \cdot 284$ | log. | $= 0 \cdot 6318495$ |
| $\frac{1}{2}(180° - [Fr + Sw]) =$ | $= 85° \, 28' \, 58'' \cdot 44$ | tan. | $= 1 \cdot 1023618$ |
| X | $= 0° \, 16' \, 16'' \cdot 03$ | tan. | $= 7 \cdot 6750415$ (44) |
| | 2 | | |
| $C = 2X$ | $= 0° \, 32' \, 32'' \cdot 06$ | | |
| $(180° - Fr) = F$ | $= 172° \, 06' \, 08'' \cdot 88$ | | |
| $180° + Sw = S$ | $= 181° \, 08' \, 12'' \cdot 00$ | | |
| $C'$ | $= 6° \, 13' \, 07'' \cdot 06$ | | |
| | $360° \, 00' \, 00'' \cdot 00$ | | |

| | | | |
|---|---|---|---|
| | $180° \, 00' \, 00'' \cdot 00$ | | |
| $C'_2$ | $= 6° \, 13' \, 07'' \cdot 06$ | | |
| | $2)173° \, 46' \, 52'' \cdot 94$ | | |
| $S_2 = F_2 = \frac{1}{2}(180° - C'_2)$ | $= 86° \, 53' \, 26'' \cdot 47$ | $S_2 =$ | $86° \, 53' \, 26'' \cdot 47$ |
| F | $= 172° \, 06' \, 08'' \cdot 88$ | $S =$ | $181° \, 08' \, 12''$ |
| $F_1$ | $= 85° \, 12' \, 42'' \cdot 41$ | $S_1 =$ | $94° \, 14' \, 45'' \cdot 53$ |
| $S_1$ | $= 94° \, 14' \, 45'' \cdot 53$ | | |
| $2X = C_1$ | $= 0° \, 32' \, 32'' \cdot 06$ | | |
| | $180° \, 00' \, 00'' \cdot 00$ | | |

| | | | |
|---|---|---|---|
| $S_1$ | $= 94° \, 14' \, 45'' \cdot 53$ | co. ar. sin. | $= 0 \cdot 0011936$ |
| $r + \frac{1}{2}h$ | $= 5731 \cdot 947$ feet | log. | $= 3 \cdot 7583021$ |
| $C_1$ | $= 0° \, 32' \, 32'' \cdot 06$ | sin. | $= 7 \cdot 9760615$ |
| $c$ | | log. | $= 1 \cdot 7355572$ (45) |

[Fig. 16.]

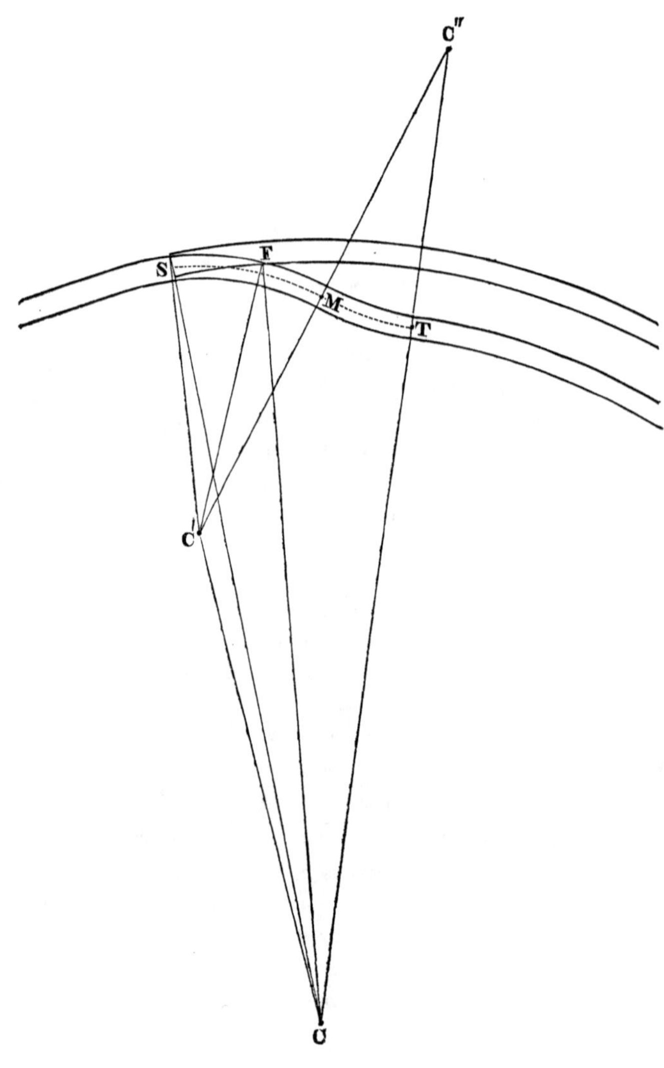

| | | | |
|---|---|---|---|
| $F_1$ | $= 85°\ 12'\ 42''\cdot41$ | co. ar. sin. | $= 0\cdot0015183$ |
| $r - \frac{1}{2}h + d$ | $= 5729\cdot663$ feet | log. | $= 3\cdot7579775$ |
| $C_1$ | $= 0°\ 32'\ 32''\cdot06$ | sin. | $= 7\cdot9760615$ |
| $c$ | $=$ | log. | $= 1\cdot7355573$ |
| $C'_2$ | $= 6°\ 13'\ 07''\cdot06$ | co. ar. sin. | $= 0\cdot9652810$ |
| $F_2$ | $= 86°\ 53'\ 26''\cdot47$ | sin. | $= 9\cdot9993601$ |
| $r'$ | $= 501\cdot41$ feet | log. | $= 2\cdot7001984$ (44) |

We have thus found $r' = 501\cdot41$ feet. It was intended as a reverse of the previous case; there we assumed $r' = 499\cdot725$; the difference is a trifle, being only $1\cdot685$ feet, which is not astonishing when we consider the acuteness of the angles we have to use in some of the triangles.

I ought not to close my remarks without an acknowledgement of my indebtedness to Mr. Percival, of Sandwich, for the manner of constructing the figure which has led us to the foregoing investigation.

**(44)** We next examine a turnout from the inside of a curve in the main track.

Retaining our former notation as far as practicable, we have in the triangle S C C′, the line S C $= r - d$; S C′ $= r'$; angle S $=$ switch angle to find the side C′ C (which we denote by $s$,) and the angles C′ and C; wherefore, $(r - d) + r' : (r - d) \backsim r' :: \tan. \frac{1}{2} (180° - S) : \tan. X = \frac{\tan. \frac{1}{2}(180° - S) . (r - d) \backsim r')}{r - d + r'}$ and $\frac{1}{2}(180° - S) + X = C'$; and $\frac{1}{2}(180° - S) - X = C$.

Then, $\sin. C' : r - d :: \sin. S : s = \frac{r - d \sin. S}{\sin. C'}$; or, we have $\sin. C : r' :: \sin. S : s = \frac{r' \sin. S}{\sin. C}$ (47)

Having found $s$, we have in the triangle C C′ C″ the side C C

$= s$, as found above; the side $C' C'' = 2r'$; and the side $C\ C'' = (r - \delta) + r'$ to find the angles, which we do by formula (41)

Having thus found the angles required, we will denote the triangle S C C′ No. 1, and represent the angles in said triangle by $S_1$, $C_1$, $C'_1$; and the triangle C C′ C″ No. 2, and represent the angles by $C_2$, $C'_2$, $C''_2$; and the triangle S C′ M No. 3, and represent the angles in said triangle by $S_3$, $C'_3$, $M_3$, and the triangle C″ M T No. 4, and shall accompany the letters denoting the angles by 4; and so on of such other triangles as may enter into our investigation in the order they are presented.

We will now proceed to find the chord S M, which we shall denote by *ch* 3. In triangle No. 3, we have $C'_3 = C'_1 - C'_2$; and $\frac{1}{2}(180° - C_3) = S_3 = M_3$; then will

$$\text{Sin. } S_3 : r' :: \text{sin. } C'_3 : ch_3 = \frac{r' \text{ sin. } C'_3}{\text{sin. } S_3} \qquad (48)$$

In triangle C″ M T = No. 4, we have $C''_4 = C''_2$; and $\frac{1}{2}(180° - C''_4) = M_4 = T_4$; then,

$$\text{Sin. } M_4 : r' :: \text{sin. } C''_4 : ch_4 = \frac{r' \text{ sin. } C''_4}{\text{sin. } M_4} \qquad (49)$$

Then, putting C F C′ for No. 5, we have $C'\ F = r' + \frac{1}{2}h$; $C\ F = r - \frac{1}{2}h$; and C C′ as found in No. 1, (which we called $s$,) to find the angles. See formula (41)

Having found the angles, the angle C′ F C, or $F_5$, will = the frog angle. Then, to ascertain the chord S F, we have in the triangle C′ F S = No. 6, $C'_1 - C_5 = C_6$; and $\frac{1}{2}(180 - C'_6) = S_6 = F_6$, and

$$\text{sin. } S_6 : r' + \tfrac{1}{2}h :: \text{sin. } C'_6 : ch_6 = \frac{(r' + \frac{1}{2}h) \cdot \text{sin. } C'_6}{\text{sin. } S_6} \qquad (50)$$

which represents the distance from the mouth of the switch of the outside rail of the turnout to the frog angle.

### EXAMPLE OF CALCULATION.

Let $r = 5729 \cdot 597$ feet; $r'$ $499 \cdot 725$ feet; $h = 4 \cdot 7$ feet; $d = 0 \cdot 416$ feet; $\delta = 11$ feet; $S_1$, or switch angle, $= 1° 08' 12''$.

$$\begin{array}{rr} & 180° \ 00' \ 00'' \\ S_1 = & 1° \ 08' \ 12'' \\ \hline & 2)\overline{178° \ 51' \ 48''} \\ \hline \frac{1}{2}(180° - S_1) = & 89° \ 25' \ 54'' \end{array}$$

| | | | |
|---|---|---|---|
| $r' - d$ | $= 5729 \cdot 181$ | | |
| $r'$ | $= 499 \cdot 725$ | | |
| $(r - d) + r' =$ | $6228 \cdot 906$ | co. ar. log. $=$ | $6 \cdot 2055882$ |
| $(r - d) \backsim r' =$ | $5229 \cdot 456$ | log. $=$ | $3 \cdot 7184566$ |
| $\frac{1}{2}(180° - S_1) =$ | $89° \ 25' \ 54''$ | tan. $=$ | $2 \cdot 0035053$ |
| X | $= 89° \ 19' \ 23''$ | tan. $=$ | $1 \cdot 9275591$ |
| $C'_1$ | $= 178° \ 45' \ 17''$ | | |
| $C_1$ | $= 0° \ 06' \ 31''$ | | |
| $S^1$ | $= 1° \ 08' \ 12''$ | | |
| | $180° \ 00' \ 00''$ | | |

(47)

| | | | | | | | |
|---|---|---|---|---|---|---|---|
| $C_1$ | $= 178° \ 45' \ 17''$ | co. ar. sin. $=$ | $1 \cdot 6628906$ | $C_1$ | $= 0° \ 06' \ 31''$ | co. ar. sin. $=$ | $2 \cdot 7222486$ |
| $r - d$ | $= 5729 \cdot 181$ feet | log. $=$ | $3 \cdot 7580926$ | $r'$ | $= 499 \cdot 725$ feet | log. $=$ | $2 \cdot 6987311$ |
| $S_1$ | $= 1° \ 08' \ 12''$ | sin. $=$ | $8 \cdot 2974820$ | | | sin. $=$ | $8 \cdot 2974820$ |
| $s_1$ | $= 5229 \cdot 532$ | log. $=$ | $3 \cdot 7184617$ | | | log. $=$ | $3 \cdot 7184617$ |

To find the elements of triangle No. 2, we have

| | | | |
|---|---|---|---|
| $a = r - \delta + r' =$ | $6218 \cdot 322$ | | |
| $b = s_1$ | $= 5229 \cdot 532$ | | |
| $c = 2r'$ | $= 999 \cdot 450$ | | |
| | $2)\overline{12447 \cdot 304}$ | | |
| $p$ | $= 6223 \cdot 652$ | log. $=$ | $3 \cdot 7940453$ |
| $p - a$ | $= 5 \cdot 330$ | log. $=$ | $0 \cdot 7267272$ |
| $p - b$ | $= 994 \cdot 120$ | log. $=$ | $2 \cdot 9974388$ |
| $p - c$ | $= 5224 \cdot 202$ | log. $=$ | $3 \cdot 7180200$ |

| | | | | | | |
|---|---|---|---|---|---|---|
| $p$ | co. ar. log. $=$ | $6 \cdot 2059547$ | | $p$ | co. ar. log. $=$ | $6 \cdot 2059547$ |
| $p - a$ | co. ar. log. $=$ | $9 \cdot 2732728$ | | $p - b$ | co. ar. log. $=$ | $7 \cdot 0025612$ |
| $p - b$ | log. $=$ | $2 \cdot 9974388$ | | $p - c$ | log. $=$ | $3 \cdot 7180200$ |
| $p - c$ | log. $=$ | $3 \cdot 7180200$ | | $p - a$ | log. $=$ | $0 \cdot 7267272$ |
| | | $2)\overline{22 \cdot 1946863}$ | | | | $2)\overline{17 \cdot 6532631}$ |
| $85° \ 25' \ 50'' \cdot 22$ | tan. $=$ | $11 \cdot 0973431$ | | $3° \ 50' \ 16'' \cdot 79$ | tan. $=$ | $8 \cdot 8266315$ |
| $2$ | | | | $2$ | | |
| $C'_2 = 170° \ 51' \ 40'' \cdot 44$ | | | | $C''_2 = 7° \ 40' \ 33'' \cdot 58$ | | |

| | | | |
|---|---|---|---|
| $p$ | co. ar. | log. | = 6·2059547 |
| $p - c$ | co. ar. | log. | = 6·2819800 |
| $p - a$ | | log. | = 0·7267272 |
| $p - b$ | | log. | = 2·9974388 |
| | | | 2)16·2121007 |
| | 0° 43′ 53″ | tan. | = 8·1060503 |
| | 2 | | |
| $C_2$ | = 1° 27′ 46″ | | |

| | |
|---|---|
| $C'_2$ | = 170° 51′ 40″·45 |
| $C''_2$ | = 7° 40′ 33″·58 |
| $C_2$ | = 1° 27′ 46″·00 |
| Proof | 180° 00′ 00″·00 |

In triangle No. 3, we have

| | | | | |
|---|---|---|---|---|
| $C'_1$ | = 178° 45′ 17″ | | | |
| $C'_2$ | = 170° 51′ 40″·44 | | | |
| $C'_3$ | = 7° 53′ 36″·56 | | | |
| | 2)172° 06′ 23″·44 | | | |
| $\frac{1}{2}$ (180 — $C'_3$) = $M_3$ | = 86° 03′ 11″·72 | co. ar. | sin. | = 0·0010312 |
| $r'$ | = 499·725 feet | | log. | = 2·6987311 |
| $C'_3$ | = 7° 53′ 36″·56 | | sin. | = 9·1377717 |
| $ch_3$ | = 68·791 feet | | log. | = 1·8375340 |

In triangle No. 4, we have

| | | | | |
|---|---|---|---|---|
| $C''_2 = C''_4$ | = 7° 40′ 33″·58 | | | |
| | 2)172° 19′ 26″·42 | | | |
| $\frac{1}{2}$ (180° — $C''_4$) = $M_4$ | = 86° 09′ 43″.21 | co. ar. | sin. | = 0·0009751 |
| $r'$ | = 499·725 feet | | log. | = 2·6987311 |
| $C_4$ | = 7° 40′ 33″·58 | | sin. | = 9·1257121 |
| $ch_4$ | = 66·899 feet | | log. | = 1·8254183 |

In triangle No. 5, we have C F $= r - \frac{1}{2} h$; C C′ $= s =$ 5229·565; C′ F $= r' + \frac{1}{2} h$, to find the angles. Let

| | | |
|---|---|---|
| $a = r - \frac{1}{2} h$ | = 5727·247 | |
| $b = r' + \frac{1}{2} h$ | = 502·075 | |
| $c = s_1$ | = 5229·565 | |
| | 2)11458·887 | |
| $p$ | = 5729·443 | log. = 3·7581123 |
| $p - a$ | = 2·196 | log. = 0·3416323 |
| $p - b$ | = 5227·368 | log. = 3·7182831 |
| $p - c$ | = 499·878 | log. = 2·6988641 |

| | | | |
|---|---|---|---|
| $p$ | co. ar. | log. = | 6·2418877 |
| $p - a$ | co. ar. | log. = | 9·6583677 |
| $p - b$ | | log. = | 3·7182831 |
| $p - c$ | | log. = | 2·6988641 |
| | | | 2)22·3174026 |
| 86° 01′ 50″·13 | | tan. = | 11·1587013 |
| 2 | | | |
| $C'_5$ = 172° 03′ 40″·26 | | | |

| | | | |
|---|---|---|---|
| $p$ | co. ar. | log. = | 6·2418877 |
| $p - b$ | co. ar. | log. = | 6·2817169 |
| $p - c$ | | log. = | 2·6988641 |
| $p - a$ | | log. = | 0·3416323 |
| | | | 2)15·5641010 |
| 0° 20′ 48″·73 | | tan. = | 7.7820505 |
| 2 | | | |
| $C_5$ = 0° 41′ 37″·46 | | | |
| $C_5$ 172° 03′ 40″·26 | | | |
| $F_5$ 7° 14′ 42″·27 | | | |
| 180° 00′ 00″·00 | | | |

| | | | |
|---|---|---|---|
| $p$ | co. ar. | log. = | 6·2418877 |
| $p - c$ | co. ar. | log. = | 7·3011359 |
| $p - a$ | | log. = | 0·3416323 |
| $p - b$ | | log. = | 3·7182831 |
| | | | 2)17·6029390 |
| 3° 37′ 21″·13 | | tan. = | 8·8014695 |
| 2 | | | |
| $F_5$ = 7° 14′ 42″·26 Frog angle. | | | |

In triangle No. 6, we have

| | | |
|---|---|---|
| $C'_1$ | = | 178° 45′ 17″ |
| $C'_5$ | = | 172° 05′ 24″·65 |
| $C'_6$ | = | 6° 39′ 52″·35 |
| | | 2)173° 20′ 07″·65 |
| $S_6 = \frac{1}{2}(180° - C'_6)$ | = | 86° 40′ 03″·82 |

| | | | |
|---|---|---|---|
| $S_6$ | = 86° 40′ 03″·82 | co. ar. sin. = | 0·0007349 |
| $r' + \frac{1}{2} h$ | = 502·075 feet | log. = | 2·7007686 |
| $C_6$ | = 6° 39′ 52″·35 | sin. = | 9·0646679 |
| $ch_6$ | = 58·233 feet | log. = | 1·7661714 |

the distance from the mouth of the switch on the outside of the turnout to the frog angle in the main track.

[FIG. 17.]

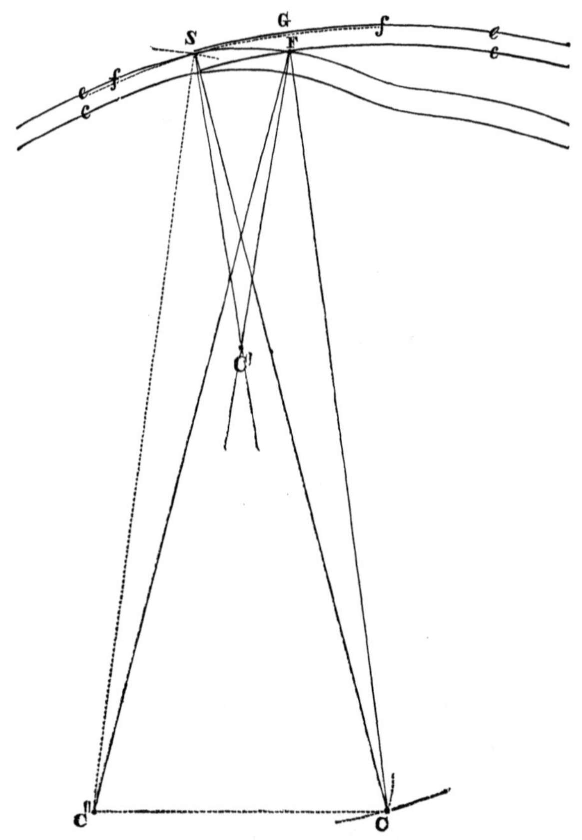

**(45)** Having thus completed our investigation of the problem direct, we will now examine it reversed, by supposing the radius of the main track given, as before, viz., $r = 5729 \cdot 597$; $h = 47$ feet; $d = 0 \cdot 416$ feet; and the frog angle $Fr = 7° \; 14' \; 42''.27$; the switch angle, $Sw = 1° \; 08' \; 12''$; to find the radius of the turnout $= r'$, and the position of the frog.

Draw the lines *ee*, *cc* representing the outer and inner rail of the main track, and the dotted line *ff*, corresponding to the switching of the outer rail; then, draw the radius F C; then, from F draw the line F C′ indefinitely, making an angle with F C = the frog angle; then, from F draw the line F C″, making an angle with F C = the switch angle + the frog angle, and equal in length to the radius of the dotted line = C G; then, with a radius = C F and with C″ as a centre, draw the angular dotted line at S, and this dotted line will intersect *ff* at the mouth of the switch. From this intersection draw the radius S C; then, draw the line S C′, making an angle with S C = the switch angle, and the lines S C′ and F C′ will intersect each other at the centre of the curve of the turnout, viz., at C′; then, with a dotted line, join S C″ and C″ C, which will complete our diagram.

If we now examine our diagram, we shall find it to contain two equal and similar triangles, viz., F C C″ and S C″ C with the angles F C″ C = S C C″; then it will be apparent that the angle S C F will be equal to the difference between the angles F C C″ and F C″ C. Having thus shown the relative magnitude of the angles last named, we will now proceed to find the angle S C F. In the triangle F C C″ we have the angle at F = the frog angle + the switch angle; the side F C $= r - \frac{1}{2} h$; the side F C″ $= r + (\frac{1}{2} h$

$- d;$) then, $(r + \frac{1}{2} h - d) + (r - \frac{1}{2} h) : (r + \frac{1}{2} h - d) \backsim (r - \frac{1}{2} h) :: \tan. \frac{1}{2} [180 - (Fr + Sw)] : \tan. X$; that is,

$$\text{Tan. } X = \frac{(r + \frac{1}{2} h - d) \backsim (r - \frac{1}{2} h) . \tan. \frac{1}{2} [180 - (Fr + Sw)]}{(r + \frac{1}{2} h - d) + (r - \frac{1}{2} h)} \qquad (51)$$

And 2 X = F C S, (51,) which we shall hereafter represent by $C_2$.

In the above notation, we have represented the frog angle by *Fr*, and the switch angle by *Sw*.

We now have the triangle F C S, which we shall hereafter denominate No. 2; the angle $C = C_2$, as found above; the line C F $= r - \frac{1}{2} h$; the line C S $= r + \frac{1}{2} h - d$, to find the angles at F and S, which we shall denote thus, by $F_2$ and $S_2$; then, $(r + \frac{1}{2} h - d) + (r - \frac{1}{2} h) : (r + \frac{1}{2} h - d) \backsim (r - \frac{1}{2} h) :: \tan. \frac{1}{2} (180 - C_2) : \tan. X_2$ and $\frac{1}{2} (180 - C_2) + X_2 = F_2$, and $\frac{1}{2} (180 - C_2) - X_2 = S_2$ (52)

Then, $\sin. F_2 : r + \frac{1}{2} h - d :: \sin. C_2 :$ S F, which we shall denote by *ch*, or $\sin. S_2 : r - \frac{1}{2} h :: \sin. C_2 : ch$. (53)

We then have in the triangle S F C′, (which we denominate No. 3, and mark the letters denoting the angles accordingly,) the line S F $= ch$, found above; the angle $S_3 = (S_2 + Sw;)$ the angle $F_3 = (F_2 - Fr) = (S_2 + Sw,)$ and the angle $C'_3 = 180 - [(F_2 - Fr) + (S_2 + Sw;)]$ and $C'_3 : ch :: F_3 : r' + \frac{1}{2} h$ (54)

Then, by subtracting $\frac{1}{2} h$, we have $r'$.

Having thus obtained our formula, we now give an example of calculation.

To find $r'$ from the frog angle,

| | | |
|---|---|---|
| $Fr$ | = | 7° 14′ 42″·27 |
| $Sw$ | = | 1° 08′ 12″· |
| $Fr + Sw$ | = | 8° 22′ 54″·27 |
| | | 2) 171° 37′ 05″·73 |
| $\frac{1}{2}$ [180° — $(Fr + Sw)$] | = | 85° 48′ 32″·86 |

| | | | | |
|---|---|---|---|---|
| $r + \frac{1}{2} h - d$ | = 5731·531 | | | |
| $r - \frac{1}{2} h$ | = 5727·247 | | | |
| Sum | = 11458·778 | co. ar. | log. = | 5·9408619 |
| Difference | = 4·284 | | log. = | 0·6318495 |
| $\frac{1}{2}$ [180°— $(Fr + Sw)$] | = 85° 48′ 32″·46 | | tan. = | 1·1350427 |
| X | = 0° 17′ 32″·38 | | tan. = | 7·7077541 |
| | 2 | | | |
| $C_2$ | = 0° 35′ 04″·76 | | | |
| | 2) 179° 24′ 55″·24 | | | |
| $\frac{1}{2}$ (180° — $C_2$) | = 89° 42′ 27″·62 | | | |
| $r + \frac{1}{2} h - d$ | = 5731·531 | | | |
| $r - \frac{1}{2} h$ | = 5727·247 | | | |
| | 11458·778 | co. ar. | log. = | 5·9408619 |
| | 4·284 | | log. = | 0·6318495 |
| $\frac{1}{2}$ (180° — $C_2$) | = 89° 42′ 27″·62 | | tan. = | 2·2922459 |
| $X_2$ | = 4° 11′ 27″·14 | | tan. = | 8·8649573 |
| $F_2$ | = 93° 53′ 54″·76 | | | |
| $S_2$ | = 85° 31′ 00″·48 | | | |
| $C_2$ | = 00° 35′ 04″·76 | | | |
| | 180° 00′ 00″·00 | | | |

| | | | | | |
|---|---|---|---|---|---|
| $F_2$ | = 93° 53′ 54″·76 co. ar. sin. = 0·0010062 | | $S_2$ | = 85° 31′ 00″·48 co. ar. sin. = 0·0013309 | |
| $r + \frac{1}{2} h - d$ | = 5731·531 feet | log. = 3·7582707 | $r - \frac{1}{2} h$ | = 5727·247 feet | log. = 3·7579458 |
| $C_2$ | = 0° 35′ 04″·76 | sin. = 8·0087699 | | | sin. = 8·0087699 |
| $ch_2$ | = 58·620 feet | log. = 1·7680468 | | | log. = 1·7680468 |

In triangle No. 3, we have

| | | | | | |
|---|---|---|---|---|---|
| $F_2$ | = 93° 53′ 54″·76 | | $S_2$ | = | 85° 31′ 00″·48 |
| $Fr$ | = 7° 14′ 42″·27 | | $Sw$ | = | 1° 08′ 12″ |
| $F_3$ | = 86° 39′ 12″·49 | | $S_3$ | = | 86° 39′ 12″·48 |
| $S_3$ | = 86° 39′ 12″·48 | | | | |
| | 173° 18′ 24″·97 | | | | |
| 180° — $(F_3 + S_3)$ | = 6° 41′ 35″·03 | | | | |

[Fig. 18.]

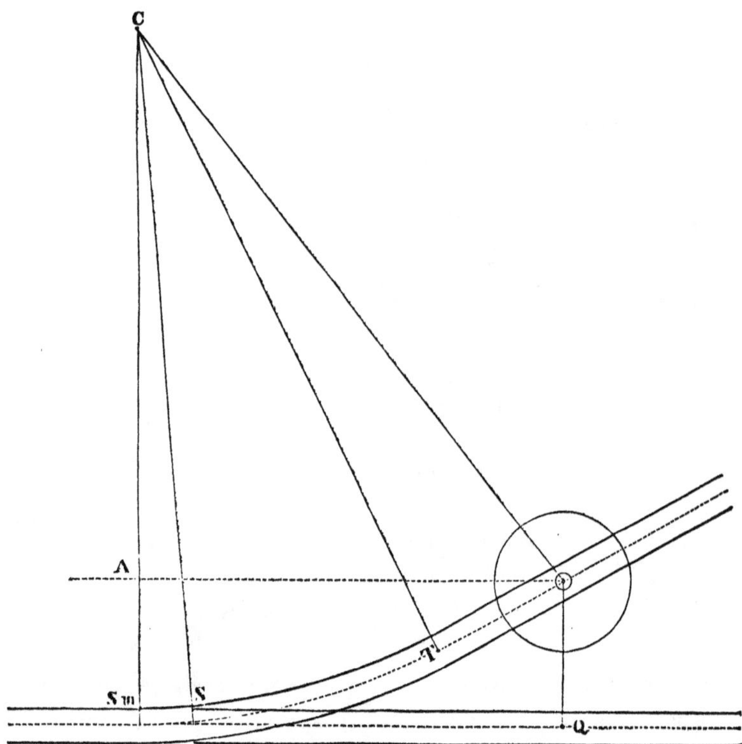

By the problem, $F_3$ and S should be equal. Then,

| | | | | |
|---|---|---|---|---|
| $180° - (F_3 + S_3)$ | = | 6° 41′ 35″·03 | co. ar. | sin. = 0·9334859 |
| $ch_2$ | = | | | log. = 1·7680467 |
| $F_3$ | = | 86° 39′ 12″·48 | | sin. = 9·9992587 |
| $r' + \frac{1}{2}h$ | = | 502·101 | | log. = 2·7007913 |
| $\frac{1}{2}h$ | = | 2·350 | | |
| $r'$ | = | 499·751 | | |

It was expected the radius would be found to be 499·725. The error only amounted to ·026 feet, or a little more than one fourth of an inch.

**(46)** The following problem has frequently presented itself in the practice of the writer, viz., the situation of a turning table with respect to the main track, (the radius of the turnout curve being given to find the relative situation of the switch,) and such additional elements as will be required to locate the turnout.

To explain: we have in several instances found it necessary so to place a turntable by the side of the railroad track, that a building erected over it might answer the purpose of shielding the table from the weather, and the engine during the night, occasionally, if not constantly; and also, to afford a convenient situation for a water tank to distribute water to the engines when upon the main track, and when sheltered upon the turntable.

The ruling principles which govern in this matter may be stated thus:

First, The proper distance between the centre of the table and the main track.

Second, A suitable amount of straight track to guide the engine steadily upon the table. (A turning table should never be directly connected with a curve, as the engine will have a tendency to force it out of place.)

Third, The remainder of the track to be united to a curve of fixed radius, which shall just connect the straight track adjoining the table with the main track.

INVESTIGATION. Representing the mouth of the switch by S; the length of the switch by *s*, and the switch angle by *Sw;* the centre of the table by ⊙; the point where the curve unites with the straight line adjoining the table by T; the centre of the curve by C; the radius by $r'$; the point on the main track where, a line being drawn therefrom to the centre of the table shall form a right angle with the centre line of said track, by Q; we shall then have in the triangle, *Sw* C S, for finding the line *Sw* C, which line we represent by $q$,

$$R : r' + d :: \cos. Sw : q = (r' + d) . \cos. Sw \qquad (55)$$

And in the triangle C ⊙ T, (by assuming C ⊙ as radius,) we have this analogy:

$$r' : T \odot :: (C \odot = \text{radius,}) : \tan. C_2 = \frac{T \odot}{r'} \qquad (56)$$

$$\text{And therefore, } \cos. C_2 : r' :: R : C \odot = \frac{r'}{\cos. C_2} \qquad (57)$$

Representing the distance of the centre of the turning table from the main track (viz., Q ⊙) by $n$, and the line C ⊙ by $p$, we then draw the line ⊙ A, parallel with the centre line of main track, and in the triangle A ⊙ C we have

$$p : R :: q - n : \cos. C = \frac{q - n}{p} \qquad (58)$$

If we now deduct from the angle C, the angles *Sw* and $C_2$, we shall have left the angle C of the triangle S C T, which we represent by $C_3$, then will $\frac{1}{2}(180 - C_3) = S = T$; and

$$\text{Sin. } T : r' :: \sin. C_3 : S\,T = \frac{r' \sin. C_3}{\sin. T} \qquad (59)$$

To give an example of calculation, we will assume $n = 36{\cdot}4$ feet; $r' = 499{\cdot}725$ feet; $s = 21$ feet; T ⊙ $= 60$ feet; $d = 0{\cdot}416$ feet; $Sw = 1° \ 08' \ 12''$. Then,

| | | | | |
|---|---|---|---|---|
| | $r' + d$ | = 500·141 | | log. = 2·6990925 |
| | $Sw$ | = 1° 08′ 12″ | | cos. = 9·9999145 |
| | $q$ | = 500·043 | | log. = 2·6990070 |
| | | | | |
| | T ⊙ | = 60 feet | | log. = 1·7781513 |
| | $r'$ | = 499·725 | co. ar. | log. = 7·3012689 |
| | $C_2$ | = 6° 50′ 47″·41 | | tan. = 9·0794202 |
| | | | | |
| | $C_2$ | = 6° 50′ 47″·41 | co. ar. | cos. = 0·0031080 |
| | $r'$ | = 499·725 | | log. = 2·6987311 |
| | $p$ | = 503·314 | | log. = 2·7018391 |
| $Sw$ = 1° 08′ 12″ | $q - n$ | = 463·643 | | log. = 2·6661837 |
| $C_2$ = 6° 50′ 47″·41 | C | = 22° 54′ 02″·78 | | cos. = 9·9643446 |
| 7° 58′ 59″·41 | $(Sw + C_2)$ | = 7° 58′ 59″·41 | | |
| | $C_3$ | = 14° 55′ 03″·37 | | |
| | | 2)165° 04′ 56″·63 | | |
| | T = ½ (180° − $C_3$) | = 82° 32′ 28″·31 | co. ar. | sin. = 0·0036904 |
| | $r'$ | = 499·725 | | log. = 2·6987311 |
| | $C_3$ | = 14° 55′ 03″·37 | | sin. = 9·4106586 |
| | S T | = 129·742 feet | | log. = 2·1130801 |

Having thus ascertained the elements deemed necessary, before we commence the business of location, we will now proceed to describe the operations necessary to execute the work.

**(47)** An examination of the figure will render it apparent that taking from the complement of $C_2$ the complement of C will leave the angle A ⊙ T. Having thus obtained A ⊙ T, we place our instrument at ⊙ and lay off said angle from A, and measure the distance $\odot$ T to T; then, moving the instrument to T, and pointing it to $\odot$, we lay off the angle $\odot$ T S $= (90° + S)$ or $(90° + T)$ and measure the distance T S to S, the place of the mouth of the switch; and if the work has been correctly prepared, we shall be the distance $d$ from the centre line of the main track, upon the side towards the turnout curve. The curve may now be further marked by deflections, agreeably to directions given in the foregoing pages.

**(48)** Having thus obtained the formula for computing the elements of a turnout to a turntable, with a given amount of straight line to guide the engine, it will frequently be found convenient to have a formula to lay out a track to a turntable from an existing joint in the rails of the main track, with a fixed radius, and a fixed position for the table. This method of proceeding will save the trouble of cutting rails, and making unnecessary joints in the main track; and another consideration will be that of affording side track room for cars to stand upon.

In the following investigation we shall preserve the notation of the preceding formula, as far as applicable.

Making $\delta =$ the distance S Q, (as measured,) and $a =$ S $\odot$, we have, by taking $a$ as radius, the following analogy:

$$\delta : n - d :: (\text{rad.}) : \tan. S = \frac{n-d}{\delta} \qquad (60)$$

wherein S will be equal to the angle at S in the triangle Q S $\odot$;

then, $$\cos. S : \delta :: R : a = \frac{\delta}{\cos. S} \qquad (61)$$

then, representing the angle S in the triangle C S $\odot$ by $S_2$, we have $90° + Sw - S = S_2$, and

$$r' + a : r' \backsim a :: \tan. \tfrac{1}{2}(180° - S_2) : \tan. X = \frac{(r' \backsim a) . \tan. \frac{1}{2}(180° - S_2)}{r' + a}$$

and $X + \frac{1}{2}(180° - S_2)$ = the angle ⊙; and $X - \frac{1}{2}(180° - S_2)$ = the angle C (62)

We then have sin. ⊙ : $r'$ :: sin. $S_2$ : C ⊙; or, sin. C : $a$ :: sin. $S_2$ : C ⊙; then, putting $p$ = C ⊙, we have

$$p : R :: r' : \cos. C_2 = \frac{r'}{p} \qquad (63)$$

$$\text{And } \cos. C_2 : r' :: \sin. C_2 : T \odot = \tan. C_2\, r' \qquad (64)$$

Then, deducting $C_2$ from C leaves $C_3$ = the angle C in the triangle S C T; and $\frac{1}{2}(180° - C_3)$ = the angle S = the angle T; and

$$\text{Sin. } T : r' :: \sin. C_3 : S\,T = \frac{\sin. C_3 \,.\, r'}{\sin. T} \qquad (65)$$

To give an example of calculation, we assume $n = 36{\cdot}4$; $r' = 499{\cdot}725$; $d = 0{\cdot}416$; $Sw = 1° \; 08' \; 12''$; $\delta = 200$ feet.

| | | | | | |
|---|---|---|---|---|---|
| $\delta$ | = 200 feet | co. ar. | log. = 7·6989700 | | 90° 00′ 00″·00 |
| $n - d$ | = 35·984 | | log. = 1·5561094 | | 1° 08′ 12″ |
| S | = 10° 11′ 58″·32 | | tan. = 9·2550794 | 90° + $Sw$ | = 91° 08′ 12″ |
| | | | | S | = 10° 11′ 58″·32 |
| S | = | co. ar. | cos. = 0·0069179 | $S_2$ | = 80° 56′ 13″·68 |
| $\delta$ | = 200 feet | | log. = 2·3010300 | | 2(99° 03′ 46″·32 |
| $a$ | = 203·211 feet | | log. = 2·3079479 | $\frac{1}{2}(180° - S_2)$ | = 49° 31′ 53″·16 |

(60)

| | | | |
|---|---|---|---|
| $r'$ | = 499·725 feet | | |
| $a$ | = 203·211 | | |
| Sum | = 702·936 | co. ar. | log. = 7·1530842 |
| Difference | = 296·514 | | log. = 2·4720452 |
| $\frac{1}{2}(180° - S_2)$ | = 49° 31′ 53″·16 | | tan. = 0·0689836 |
| X | = 26° 18′ 34″·57 | | tan. = 9·6941130 |
| ⊙ | = 75° 50′ 27″·73 | | |
| C | = 23° 13′ 18″·59 | | |
| $S_2$ | = 80° 56′ 13″·68 | | |
| | 180° 00′ 00″·00 | | |

| | | | | | |
|---|---|---|---|---|---|
| $\odot$ | $= 75° 50' 27''·73$ | co. ar. sin. $= 0·0133981$ | $C = 23° 13' 18''·59$ | co. ar. sin. $= 0·4041819$ | |
| $r'$ | $= 499·725$ feet | log. $= 2·6987311$ | $a = 203·211$ feet | log. $= 2·3079472$ | |
| $S_2$ | $= 80° 56' 13''·68$ | sin. $= 9·9945442$ | | sin. $= 9·9945442$ | |
| $p$ | $=$ | log. $= 2·7066734$ | | log. $= 2·7066733$ | |
| $r'$ | $=$ | log. $= 2·6987311$ | | | |
| $C_2$ | $= 10° 55' 27''·50$ | cos. $= 9·9920577$ | $C = 23° 13' 18''·59$ | | (63) |
| | | | $C_2 = 10° 55' 27''·50$ | | |
| $C_2$ | $= 10° 55' 27''·50$ | tan. $= 9·2855789$ | $C_3 = 12° 17' 51''·09$ | | |
| $r'$ | $=$ | log. $= 2·6987311$ | $2)167° 42' 08''·91$ | | |
| $T \odot$ | $= 96·452$ feet | log. $= 1·9843100$ | $\frac{1}{2}(180° - C_3) = 83° 51' 04''·45$ | | (64) |

| | | | |
|---|---|---|---|
| $T$ | $= 83° 51' 04''·45$ | co. ar. sin. $= 0·0025057$ | |
| $r'$ | $= 499·725$ feet | log. $= 2·6987311$ | |
| $C_3$ | $= 12° 17' 51''·09$ | sin. $= 9·3283555$ | |
| $S\ T$ | $= 107·051$ feet | log. $= 2·0295923$ | (65) |

Having thus ascertained the elements of the turnout, it remains to describe the method of locating or marking the same upon the field.

We have found, formula (63,) $C_2 = 10° 55' 27''·5$, the complement to which $= 79° 04' 32''·5 =$ the angle $C \odot T$; we have also found, formula (62,) $\odot = 75° 50' 27''·73 =$ the angle $C \odot S$.

We now place our instrument at $\odot$, and lay off from S the difference between $79° 04' 32''·5$ and $75° 50' 27''·73 = 3° 14' 04''·77$, and measure from $\odot$ to T 96·452 feet; we then move the instrument to T, and lay off the angle $S\ T \odot = \frac{1}{2}(180° - C_3) + 90°$ that is $= 90° + T$, as found above (65) $= 90° + 83° 51' 04''·45 = 173° 51' 04''·45$, and measure 107·051 feet to S; and if the field work and computations have been correctly performed, the point *s* will be found directly between the joints in the rails in the main track, and 0·416 feet from the centre line on the side of the turnout. The curve may then be further marked by deflections, as heretofore explained.

**(49)** It is not an unfrequent occurrence for an engineer to be required to locate railroad and other curves in situations inaccessible to the making of measurements in the common or ordinary methods. These cases occur where railroads are located across bays and inlets of the ocean or lakes, and across rivers or estuaries, etc., etc. We know of no better method of managing this matter than by projecting a system of triangles from a well-selected base to points desirable to mark or permanently fix.* The calculations necessary for the arrangement of such a series of triangles, when connected with the choice of the location of the curve, and the determination of the necessary elements to carry forward the whole work with accuracy and convenience, may, in some instances, be too complicated for the invention of the young and inexperienced engineer. To aid such in the performance of their task is the object of the present article.

We have chosen as an example, an imaginary river of some 250 feet wide; but, before we proceed with the investigations, let us suppose the straight or tangent lines upon both sides of the river to have been located, and sufficiently marked to show their relative bearings. Our first operation, then, is to select such a situation as may be thought, upon a thorough examination of the whole subject, the most desirable for the location of the curve. The point we have selected will be seen in Fig. 19, marked 1; and in Fig. 20, marked as station 10 of the railroad location.

Having determined on this point, we commence the survey by running a line from it to a point in the tangent line upon the same

* Care should be taken to so select the termini of the base, that the lines projected therefrom should intersect with each other at the points to be located as near at right angles as they well can.

[Fig. 19.]

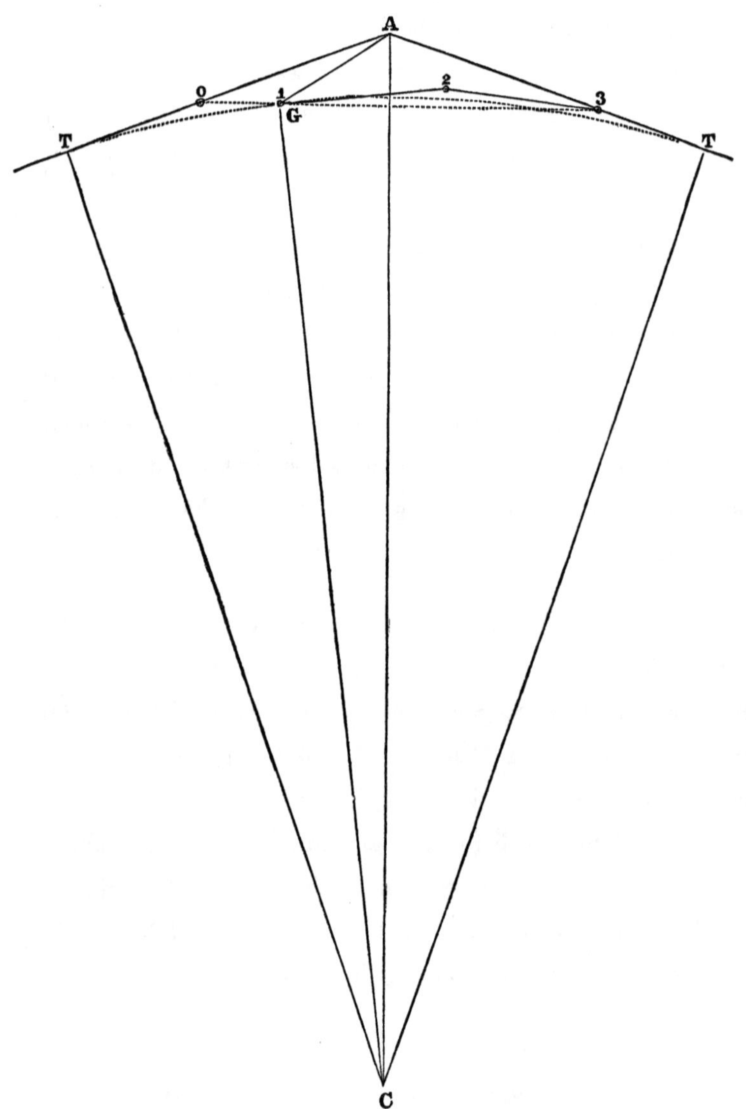

side of the river, marked 0 on the diagram. Then, by considering the tangent line from 0 to A to bear due east, (whatever may be its direction,) we measure the angle A 0 1, by which we determine the line from 0 to 1 to bear S.E. 75° 08′ 35″·37, and by measurement we find the distance 152·5 feet. We then place a signal at station 2, and by a triangulation we ascertain the bearing from station 1 to 2 to be S.E. 85° 02′ 37″, and distance = 350 feet; then, crossing the river to station 2, we run and measure a line therefrom to station 3, situated in the other tangent line; the bearing of this line we determined to be S. 72° 59′ 36″·41 E., and its length = 1604·264 feet; then, removing to station 3, we ascertain the bearing of the tangent line to be S. 60° E., or (which is the same,) N. 60° W.

Having thus connected the straight or tangent lines, by a traverse running through the point selected as the most suitable for the location of the curve, our next step will be to prepare our data to ascertain its radius.

Upon examination of the foregoing, we find the courses and distances noted in the following table, viz.,

| | BEARINGS. | DISTANCES. | |
|---|---|---|---|
| | ° ′ ″ | | |
| Station 0 to Station 1 = | S. 75 08 35·37 E. = | 152·5 | feet |
| " 1 " 2 = | S. 85 02 37·00 E. = | 350·0 | " |
| " 2 " 3 = | S. 72 59 36·41 E. = | 1604·264 | " |

and the bearing of the tangent lines from 0 to A due East, and from A to 3 S. 60° E.

Having thus collected and arranged our courses and distances, we then compute their northings and southings, eastings and westings, according to the requirements of the case.

Representing the northing by N, southing by S, easting by E, and westing by W; and, for the convenience of a general expression in our formula, we call the northings or southings the latitudes, which we represent by L; and the eastings and westings the departures, which we represent by D. Then, putting R for the radius of the tables, B for the bearings, and $d$ for the distance, we have

$$\text{R} : d :: \sin. \text{B} : \text{D} = \sin. \text{B} . d;$$

$$\text{and R} : d :: \cos. \text{B} : \text{L} = d . \cos. \text{B} \qquad \text{(M)}$$

COMPUTATIONS OF LATITUDES AND DEPARTURES.

| | | | | | |
|---|---|---|---|---|---|
| No. 1. | B = S 75° 08′ 35″·37 E | sin. = 9·9852331 | | | cos. = 9·4089262 |
| | $d$ = 152·5 feet | log. = 2·1832698 | | | log. = 2·1832698 |
| | D = 147·401 feet | log. = 2·1685029 | L = | 39·1017 | log. = 1·5921960 |
| No. 2. | B = S 85° 02′ 37″ E | sin. = 9·9983730 | | | cos. = 8·9365008 |
| | $d$ = 350 feet | log. = 2·5440680 | | | log. = 2·5440680 |
| | D = 348·691 feet | log. = 2·5424410 | L = | 30·239 | log. = 1·4805688 |
| No. 3. | B = S 72° 59′ 36″·41 E | sin. = 9·9805811 | | | cos. = 9·4660977 |
| | $d$ = 1604·264 feet | log. = 3·2052779 | | | log. = 3·2052779 |
| | D = 1534·119 feet | log. = 3·1858590 | L = | 469·219 | log. = 2·6713756 |

Having computed the latitudes and departures, or, in other words, the southings and eastings indicated by the tables of courses and distances; we then, to render these operations as perspicuous as we well can, re-arrange in a tabular form, our courses and distances, with the southings and eastings belonging to each; and having summed them up, we proceed to compute the bearing and distance from 0 to station 3. Thus, by making use of the symbols of the preceding formula, with the addition of $\delta$, by which we represent the distance from station 0 to 3, we have

$$\text{Tan. B} = \frac{\text{D}}{\text{L}}; \text{ and sin. B} : \text{D} :: \text{R} : \delta = \frac{\text{D}}{\sin. \text{B}};$$

$$\text{or, Cos. B} : \text{L} :: \text{R} : \delta = \frac{\text{L}}{\cos. \text{B}} \qquad \text{(N)}$$

| | Bearings. | Distances. | Southings. | Eastings. |
|---|---|---|---|---|
| Station 0 to 1 | S 75° 08′ 35″·07 E | 152·5 feet | 39·102 | 147·401 |
| " 1 to 2 | S 85° 02′ 37″·00 E | 350·0 " | 30·239 | 348·691 |
| " 2 to 3 | S 72° 59′ 36″·41 E | 1604·264 " | 469·219 | 1534·119 |
| | | | L = 538·560 | D = 2030·211 |

D = sum of eastings = 2030·211 log. = 3·3075411

L = sum of southings = 538·560 log. = 2·7312341

Bearing from station 0 to 3 = B = S 75° 08′ 35″·37 E tan. = 0·5763070

B = 75° 08′ 35″·37 sin. = 9·9852331 cos. = 9·4089261

D = 2030·211 log. = 3·3075411 L = 538·560 log. = 2·7312341

$\delta$ = 2100·429 feet log. = 3·3223080 Proof log. = 3·3223080

Having thus obtained the bearing from station 0 to 3, viz., S 75° 08′ 35″·37 E, and distance = 2100·429 feet, our next step will be to ascertain the distances 0 A and A 3. In the triangle A 3 0, we have to find the several angles. The bearing from

0 to A = due East
0 to 3 = S 75° 08′ 35″·37 E } which gives angle at 0 = 14° 51′ 24″·63

A to 0 = due West
A to 3 = S 60° 00′ 00″·00 E } " " " A = 150° 00′ 0″·00

3 to A = N 60° 00′ 00″·00 W
3 to 0 = N 75° 08′ 35″·37 W } " " " 3 = 15° 08′ 35″·37

Having found the angles, we have

$$\text{Sin. A} : \delta :: \text{sin. } 0 : \text{A } 3 = \frac{\delta \,.\, \text{sin. } 0}{\text{sin. A}} ;$$

$$\text{and Sin. A} : \delta :: \text{sin. } 3 : 0\text{ A} = \frac{\delta \,.\, \text{sin. } 3}{\text{sin. A}} \quad \text{Thus,}$$

| | | | | | |
|---|---|---|---|---|---|
| A = 150° 00′ 00″·00 co. ar. sin. = 0·3010300 | | A = | | co. ar. sin. = 0·3010300 | |
| 0 = 14° 51′ 24″·63. | sin. = 9·4089261 | δ = | | log. = 3·3223080 | |
| δ = | log. = 3·3223080 | 3 = 15° 08′ 35″·37 | | sin. = 9·4170259 | |
| A 3 = 1077·12 feet | log. = 3·0322641 | A 0 = 1097·4 feet | | log. = 3·0403639 | |

For the purpose of ascertaining the angle G in the triangle A G C, we assume a radius = unity, which we represent in our formula by 1, retaining $r$ as the radius in the unit of measure. Then, putting A for the angle at apex; T and T′ for the tangent points; C for the angle at the centre of the curve; G for the angle at station 1 in the traverse; A′ for the angle T′ A G; C′ for the angle A C G; we have in the triangle A 0 G, the angle 0 and the sides 0 A and 0 1, to find the side A 1 = A G, and the angles A′ and G. Putting $a$ for the side 0 A, and $b$ for the side 0 1 = 0 G, we have

$$a + b : a \backsim b :: \tan. \tfrac{1}{2}(180° - 0) : \tan. \tfrac{1}{2}(A' \backsim G) = \frac{(a \backsim b) . \tan. \frac{1}{2}(180° - 0)}{a + b};$$

$$\text{and } \tfrac{1}{2}(180° - 0) + \tfrac{1}{2}(A' \backsim G) = G;$$

$$\text{and } \tfrac{1}{2}(180° - 0) - \tfrac{1}{2}(A' \backsim G) = A' \qquad \text{(O)}$$

Having found the angles A′ and G, we find the side A G, which we represent by $m$, by either of the analogies following:

$$\text{Sin. } A' : b :: \sin. 0 : m = \frac{b . \sin. 0}{\sin. A'};$$

$$\text{or, Sin. } G : a :: \sin. 0 : m = \frac{a . \sin. 0}{\sin. G} \qquad \text{(P)}$$

Having obtained $m$, and putting G′ for the angle G in the triangle A C G, we have

$$\text{Sin. } \tfrac{1}{2} A : 1 :: R : A C = \frac{1}{\sin. \frac{1}{2} A};$$

then, representing A C by $n$, we have

$$1 : \sin. (\tfrac{1}{2} A - A') :: n : \sin. G' = n . \sin. (\tfrac{1}{2} A - A') \qquad \text{(Q)}$$

$$\text{And } \sin. (\tfrac{1}{2} A - A' + G') : m :: \sin. (\tfrac{1}{2} A - A') : r = \frac{m . \sin. (\frac{1}{2} A - A')}{\sin. (\frac{1}{2} A - A' + G')} \qquad \text{(R)}$$

We now introduce an example of computation [formulæ (O) and (P.)]

| | | | | | |
|---|---|---|---|---|---|
| $a$ | = 1097·4 feet | | | | 180° 00′ 00″·00 |
| $b$ | = 152·5 " | | | 0 | = 14° 51′ 24″·63 |
| $a + b$ | = 1249·9 | co. ar. log. = 6·9031247 | | | 2)165° 08′ 35″·37 |
| $a \backsim b$ | = 944·9 | log. = 2·9753858 | | ½ (180° — 0) | = 82° 34′ 17″·68 |
| ½ (180° — 0) | = 82° 34′ 17″·68 | tan. = 0·8847831 | | | |
| ½ (A′ ∽ G) | = 80° 12′ 52″·64 | tan. = 0·7632936 | | | |
| G | = 162° 47′ 10″·32 | G = 162° 47′ 10″·32 | co. ar. sin. = 0·5287991 | | |
| A′ | = 2° 21′ 25″·04 | $a$ = 1097·4 feet | log. = 3·0403650 | | |
| 0 | = 14° 51′ 24″·63 | 0 = 14° 51′ 24″·63 | sin. = 9·4089261 | | |
| Proof | = 180° 00′ 00″·00 | $m$ = 950·8 feet | log. = 2·9780902 | | |

By formulæ (Q) and (R) we have

| | | | |
|---|---|---|---|
| ½ A | = 75° 00′ 00″·00 | co. ar. | sin. = 0·0150562 = $n$ |
| ½ A — A′ | = 72° 38′ 34″·96 | | sin. = 9·9797599 |
| Ambiguous G′ | = 81° 09′ 53″·75 * | | sin. = 9·9948161 |
| True G′ | = 98° 50′ 06″·25 | | |
| ½ A — A′ | = 72° 38′ 34″.96 | | |
| ½ A — A + G′ | = 171° 28′ 41″·21 | co. ar. | sin. = 0·8291893 |
| $m$ | = 950·8 feet | | log. = 2·9780892 |
| ½ A — A′ | = 72° 38′ 34″·96 | | sin. = 9·9797599 |
| $r$ | = 6124·05 feet | | log. = 3·7870384 |

Having ascertained the radius which the problem requires, we proceed to ascertain the deflection for a chord of 50 feet.

By formula (3) we have sin. D $= \frac{\frac{1}{2} ch}{r}$ ; hence

| | | | |
|---|---|---|---|
| $r$ | = 6124·05 feet | co. ar. | log. = 6·2129616 |
| ½ $ch$ | = 25 " | | log. = 1·3979400 |
| D | = 0° 14′ 02″ | | sin. = 7·6109016 |

* As the problem under all its forms requires G′ to be larger than a right angle, it is evident that the true G′ must be the supplemental angle, inasmuch as the sine of an angle is the same as the sine of its supplement.

The deflection thus found being an awkward sum to add or subtract in the field, we may assume one more convenient without materially changing the location of the curve; we therefore assume 0° 14′ as the measure of a deflection; then, by formula (5) we have $r = \frac{\frac{1}{2} ch}{\text{sin. D}}$ ; and, for the purpose of ascertaining by what amount this change in the length of the radius will affect the location of the curve, we will endeavor to find the distance from the apex to the middle of the curve for each radius. By formula (6) we have $t = \text{tan. } \frac{1}{2} \text{ C} . r;$ and by (7) we have

$$b = t . \text{ tan. } \tfrac{1}{4} \text{ C} = \text{tan. } \tfrac{1}{2} \text{ C} . \text{ tan. } \tfrac{1}{4} \text{ C} . r = \frac{\text{tan. } \frac{1}{2} \text{ C} . \text{ tan. } \frac{1}{4} \text{ C} . \frac{1}{2} ch}{\text{sin. D}}$$

| The value of $b$, computed from the radius already obtained. | | | The value of $b$, computed to correspond to a radius based upon a deflection of 0° 14′. | | |
|---|---|---|---|---|---|
| $\frac{1}{2}$ C | = 15° 00′ 00″ | tan. = 9·4280525 | D | = 0° 14′ 00″ | co. ar. sin. = 2·3901470 |
| $\frac{1}{4}$ C | = 7° 30′ 00″ | tan. = 9·1194291 | $\frac{1}{2}$ C | = 15° 00′ 00″ | tan. = 9·4280525 |
| $r$ | = 6124·05 | log. = 3·7870384 | $\frac{1}{4}$ C | = 7° 30′ 00″ | tan. = 9·1194291 |
| $b$ | = 216·033 feet | log. = 2·3345200 | $\frac{1}{2}$ $ch$ | = 25 feet | log. = 1·3979400 |
| | | | $b$ | = 216·555 feet | log. = 2·3355686 |
| $b$ | = . . . . . . . . . . . . . . | | | 216·033 " | |
| | | | | 0·522 " | |

Thus we see that the proposed change in the deflection will affect the location of the curve only 0·522 feet, an amount in most cases too small to produce any practical inconvenience.

Having shown that whenever convenience requires a change of a few seconds in the angle of deflection, the change may be made without materially affecting the location of the curve, we now proceed to determine a radius which shall correspond with the desired deflection, viz., of 0° 14′, as explained in the foregoing. By formula (5) we have $r = \frac{\frac{1}{2} ch}{\text{sin. D}}$ Thus,

| | | |
|---|---|---|
| D | = 0° 14′ | co. ar. sin. = 2·3901470 |
| $\frac{1}{2}$ $ch$ | = 25 feet | log. = 1·3979400 |
| $r$ | = 6138·853 feet | log. = 3·7880870 |

To find the position of the tangent points at stations T and T′ we compute their distance from apex, and compare them with the distances of 0 and 3, which have been already determined. By formula (6) we have $t = \tan. \frac{1}{2} C . r$. Thus,

| | | | |
|---|---|---|---|
| $r$ | = 6138·853 feet | log. | = 3·7880870 |
| $\frac{1}{2}$ C | = 15° 00′ 00″ | tan. | = 9·4280525 |
| $t$ | = 1644·90 | log. | = 3·2161395 |
| A to 3, heretofore computed | = 1077·12 | | |
| | 567·78 | | |

We thus find the point T 567·78 feet further from A than the point 3. Again,

| | | |
|---|---|---|
| $t$ | = 1644·90 | feet |
| A to 0, heretofore computed | = 1097·40 | " |
| | 547·50 | " |

We thus find the point T′ 547·50 feet further from A than the point 0.

To find the length of arc from point T′ to point G, corresponding to point 1 in the traverse, we have in the triangle A C G, the angle at C = 180° — (the angle G′ + angle A.) Thus, we find

| | |
|---|---|
| A = ($\frac{1}{2}$ A — A′) = 75° — 2° 21′ 25″·04 | = 72° 38′ 34″·96 |
| G′ | = 98° 50′ 06″·25 |
| C = the supplement | = 8° 31′ 18″·79 |
| Proof | 180° 00′ 00″·00 |

Then, we have half the centre angle = $\frac{1}{2}$ C = 15° 00′ 00″ minus the supplementary angle C found above = the angle C in the triangle G C T′. Thus,

| | |
|---|---|
| $\frac{1}{2}$ C | = 15° 00′ 00″·00 |
| Supplementary C | = 8° 31′ 18″·79 |
| Hence the C sought | 6° 28′ 41″·21 |

[Fig. 20.]

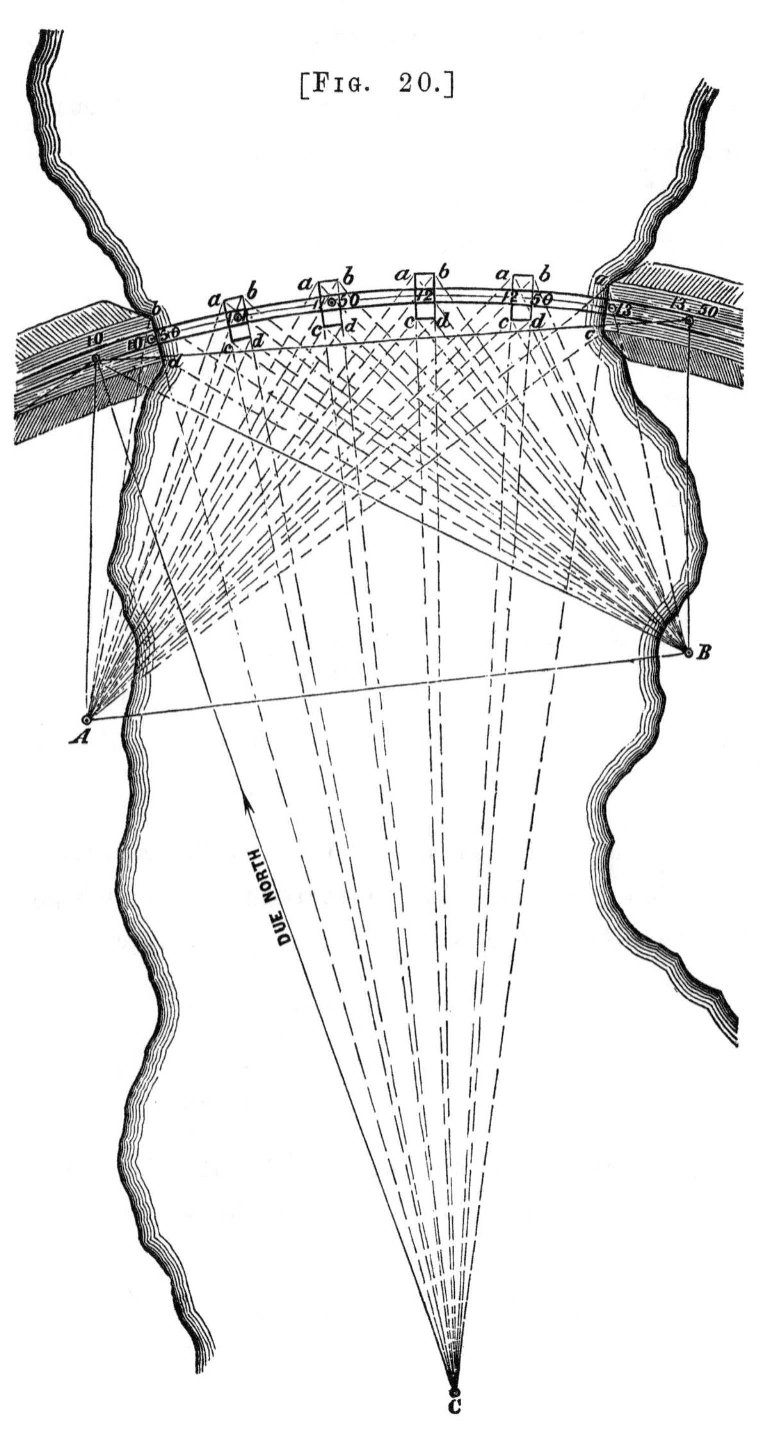

Then, representing the angle C thus found and reduced to seconds by C″, and the arc sought by $a$, we have, by formula (8,) $a = \frac{r \cdot C''}{r''}$

| | | | | |
|---|---|---|---|---|
| $C''$ | = | 23321·21 | | log. = 4·3677511 |
| $r$ | = | | | log. = 3·7880870 |
| $r''$ | = | | co. ar. | log. = 4·6855749 |
| $a$ | = | 694·09 feet | | log. = 2·8414130 |

If we now consider T′ as numbered 3·06, in the stations of location, and the numbers in the location to be increasing as we enter the curve, we shall find the point at G, or rather near G, (as we have slightly changed the radius,) to be equal to 3·06 + 6·94, which increases the number of the locating stations to 10. For the purpose of avoiding fractions, we ascertain the point T′ by measurement from 0, and then locate the curve to station 10; from this point we ascertain the direction of the radius, and select the point A, which should be so situated as to command a distinct view of the locality where the work is to be laid out. We then discover the relative direction and length of the line 10 A, by ascertaining the angle which it makes with the radius of the curve, and measuring the distance between the termini. Let us now suppose the angle to measure 20° 00′, and the length of the line to be 200 feet. We next cross the river or bay, and select the point B, which should likewise command a distinct view of the locality where the work is to be laid out; we then measure the angles which form the triangle A B 10, and compute their relative positions. Now, supposing the angles at

| | | |
|---|---|---|
| A | = | 83° 00′ |
| B | = | 32° 00′ |
| 10 | = | 65° 00′ |
| | | 180° 00′ |

Then, putting $\delta$ for the distance between 10 and A, we have Sin. B : $\delta$ :: sin. 10 : A B; and sin. B : $\delta$ :: sin. A : 10 B (S)

Thus,

| | | | | |
|---|---|---|---|---|
| B | = 32° 00′ | co. ar. | sin. = | 0·2757903 |
| $\delta$ | = 200 feet | | log. = | 2·3010300 |
| 10 | = 65° 00′ | | sin. = | 9·9572757 |
| A B | = 342·05 feet | | log. = | 2·5340960 |

| | | | |
|---|---|---|---|
| $\frac{\delta}{\text{Sin. B}}$ = | | log. = | 2·5768203 |
| A | = 83° 00′ | sin. = | 9·9967507 |
| 10 B | = 374·61 feet | log. = | 2·5735710 |

**(50)** We have, from station 10 to 10·50, a distance of 50 feet of arc; then, from 10·50 to 11, a like distance; and so on, from station to station, to station 13·50. (See Fig. 20.) To compute in the readiest manner the relative position of these several points, or rather the relative positions of the points *a b c* and *d*, (which represent the corners of the piers situated about these stations,) we assume the radius of the curve from station 10 to bear due south, and C as a zero point. Then, by ascertaining the distances and relative bearings to each of the points, we compute what we shall (for the want of more appropriate terms) call their northings or southings, eastings or westings, without consideration of their astronomical or geodetical bearings. To ascertain the angles of the radii from C to these several points, we put $a$ for the arc connecting them, $r$ for the radius in the unit of measure, and $r''$ for an arc in seconds equal in length to radius. We then have, by formula (9,) $C'' = \frac{r'' a}{r}$ Thus the angle at C, between stations 10 and 10·50, gives $a = 50$ feet. Then,

| | | | |
|---|---|---|---|
| $a$ | = 50 feet | log. = | 1·6989700 |
| $r''$ | = | log. = | 5·3144251 |
| $r$ | = | co. ar. log. = | 6·2119130 |
| $C''$ | = 1680″ | = nearly log. = | 3·2253081 |

and 1680″, reduced to degrees and minutes, will give $= 0° 28' 00''$. We have given this computation for the purpose of explaining a general rule which will apply in all cases.

In the present case, the arc $a =$ the chord of one of our deflections; and, as the difference between the chord of fifty feet in length and the arc it spans (based upon a radius of 6138·853 feet) is so small that the one may be taken for the other, in the practical operations of locating a railroad, we therefore may, without sensible error, take the angle at C for 50 feet of arc $=$ two deflections $= 28'$, or the same as above.

But we shall, notwithstanding, when we come to consider the dimensions of the piers, find a necessity for the formulæ. Let us assume the foundations of the piers to be 8 feet broad and 18 feet in length. Now, as the stations named above correspond to the centre of these piers, we find it necessary to determine half the angular width of them from C. Thus,

| | | | |
|---|---|---|---|
| $a$ | $=$ 4 feet | | log. $=$ 0·6020600 |
| $r''$ | $=$ | | log. $=$ 5·3144251 |
| $r$ | | co. ar. | log. $=$ 6·2119130 |
| $C''$ | $=$ 134″·4 | | log. $=$ 2.1283981 |

Reducing $C''$ to degrees and minutes, gives us the angle $=$ $0° 2' 14''·4$; but, for the purpose of avoiding (in the computations) the fractions of a second, we may, without varying the dimensions of the piers perceptibly, assume the angular width of the half pier to be $0° 2' 15''$. For like reasons, with a radius of the length we have adopted, we assume both ends of the pier to be of the same angular width.

Having made these explanations, we proceed to construct a table of the angular positions of the corners of the piers represented by *a b c d.* We have stated above that the centres of the piers are two deflections, or 28′, apart from C. Then, taking station 10 for a starting point, and the radius from point C through this point, as bearing due north, we have the angle to station 10·50 = 28′; and the angle to station 11, twice 28′; and so on. Having determined the position of the primitive stations, we may, by additions and subtractions of the angular half widths and widths of the piers, determine the angular positions of the points *a b c d;* and upon these principles we construct the following table, viz.,

| | Relative Bearings of Primitive Station N.E. | Relative Bearings of *b* and *d* N.E. | Relative Bearings of *a* and *c* N.E. |
|---|---|---|---|
| Station 10 to 10·50 angle = | 28′ + 2′ 15″ = | 30′ 15″ | |
| " " 11·00 " = | 56′ + " = | 58′ 15″ — 4′ 30″ = | 53′ 45″ |
| " " 11·50 " = | 84′ + " = | 1° 26′ 15″ — " = | 1° 21′ 45″ |
| " " 12·00 " = | 112′ + " = | 1° 54′ 15″ — " = | 1° 49′ 45″ |
| " " 12·50 " = | 140′ + " = | 2° 22′ 15″ — " = | 2° 17′ 45″ |
| " " 13·00 " = | 168′ + " = | " = | 2° 45′ 45″ |
| " " 13·50 " = | 196′ = N.E. 3° 16′ | | |

Having thus arranged a table of bearings from the centre of the curve, or C, of *a b c d,* with the primitive station to which they are connected, we next prepare a table containing both bearings and distances, leaving a space for the northings and eastings to be added after computation.

## A TABLE OF BEARINGS, DISTANCES, NORTHINGS AND EASTINGS, OF STATIONS, FROM C, OR THE CENTRE OF THE CURVE.

| Primitive Stations. | Bearings. | Distances. | Northings in feet. | Eastings in feet. |
|---|---|---|---|---|
| 10·00 from C to 10 | Due North | $r + 0$ feet = 6138·85 feet | 6138·85 | 00·000 |
| 10·50 " " " *d* | N.E. = 0° 30′ 15″ | $r - 9$ " = 6129·85 " | 6129·61 | 53·938 |
| " " " " *b* | " " " | $r + 9$ " = 6147·85 " | 6147·61 | 54·096 |
| 11·00 " " " *c* | N.E. = 0° 53′ 45″ | $r - 9$ " = 6129·85 " | 6129·10 | 95·838 |
| " " " " *a* | " " " | $r + 9$ " = 6147·85 " | 6147·10 | 96·119 |
| " " " " *d* | N.E. = 0° 58′ 45″ | $r - 9$ " = 6129·85 " | 6128·97 | 103·860 |
| " " " " *b* | " " " | $r + 9$ " = 6147·85 " | 6146·97 | 104·165 |
| 11·50 " " " *c* | N.E. = 1° 21′ 45″ | $r - 9$ " = 6129·85 " | 6128·12 | 145·755 |
| " " " " *a* | " " " | $r + 9$ " = 6147·85 " | 6146·11 | 146·183 |
| " " " " *d* | N.E. = 1° 26′ 45″ | $r - 9$ " = 6129·85 " | 6127·92 | 153·176 |
| " " " " *b* | " " " | $r + 9$ " = 6147·85 " | 6145·92 | 154·228 |
| 12·00 " " " *c* | N.E. = 1° 49′ 45″ | $r - 9$ " = 6129·85 " | 6126·73 | 195·662 |
| " " " " *a* | " " " | $r + 9$ " = 6147·85 " | 6144·72 | 196·237 |
| " " " " *d* | N.E. = 1° 54′ 15″ | $r - 9$ " = 6129·85 " | 6126·46 | 203·682 |
| " " " " *b* | " " " | $r + 9$ " = 6147·85 " | 6144·45 | 204·280 |
| 12·50 " " " *c* | N.E. = 2° 17′ 45″ | $r - 9$ " = 6129·85 " | 6124·93 | 245·557 |
| " " " " *a* | " " " | $r + 9$ " = 6147·85 " | 6142·92 | 246·177 |
| " " " " *d* | N.E. = 2° 22′ 15″ | $r - 9$ " = 6129·85 " | 6124·60 | 253·574 |
| " " " " *b* | " " " | $r + 9$ " = 6147·85 " | 6142·59 | 254·318 |
| 13·00 " " " *c* | N.E. = 2° 45′ 45″ | $r - 9$ " = 6129·85 " | 6122·81 | 295·435 |
| " " " " *a* | " " " | $r + 9$ " = 6147·85 " | 6140·79 | 296·302 |
| 13·50 " " " ⊙ | N.E. = 3° 16′ 00″ | $r - 0$ " = 6138·85 " | 6128·87 | 349·811 |
| Point A from C | | | 5950·91 | — 68.404 |
| " B " " | | | 5873·97 | 269·884 |

Having prepared our table of bearings and distances from the centre C to the corners of the piers, we introduce examples of computation of northings and eastings; all the bearings the table contains being north-east. Substituting $\delta$ for $d$, we have, by formula (M,)

$$D = \sin.\ B\ \delta;\ \text{and}\ L = \cos.\ B\ \delta.$$

| | | | | | |
|---|---|---|---|---|---|
| 10·50 | $d$ = N.E. | 0° 30′ 15″ | sin. = 7·9444459 | | cos. = 9·9999832 |
| | $\delta$ = | 6129·85 | log. = 3·7874499 | | log. = 3·7874499 |
| | D = | 53·938 | log. = 1·7318958 | L = 6129·61 | log. = 3·7874331 |
| 10·50 | $b$ = N.E. | | sin. = 7·9444459 | | cos. = 9·9999832 |
| | $\delta$ = | 6147·85 | log. = 3·7887233 | | log. = 3·7887233 |
| | D = | 54·096 | log. = 1·7331692 | L = 6147·61 | log. = 3·7887065 |
| 11·00 | $c$ = N.E. | 0° 53′ 45″ | sin. = 8·1940869 | | cos. = 9·9999469 |
| | $\delta$ = | 6129·85 | log. = 3·7874499 | | log. = 3·7874499 |
| | D = | 95·838 | log. = 1·9815368 | L = 6129·10 | log. = 3·7873968 |
| 11·00 | $a$ = N.E. | 0° 53′ 45″ | sin. = 8·1940869 | | cos. = 9·9999469 |
| | $\delta$ = | 6147·85 | log. = 3·7887233 | | log. = 3·7887233 |
| | D = | 96·119 | log. = 1·9828102 | L = 6147·10 | log. = 3·7886702 |
| 11·00 | $d$ = N.E. | 0° 58′ 15″ | sin. = 8·2290013 | | cos. = 9·9999377 |
| | $\delta$ = | 6129·85 | log. = 3·7874499 | | log. = 3·7874499 |
| | D = | 103·86 | log. = 2·0164512 | L = 6128·97 | log. = 3·7873876 |
| 11·00 | $b$ = N.E. | 0° 58′ 15″ | sin. = 8·2290013 | | cos. = 9·9999377 |
| | $\delta$ = | 6147·85 | log. = 3·7887233 | | log. = 3·7887233 |
| | D = | 104·165 | log. = 2·0177246 | L = 6146·97 | log. = 3·7886610 |
| 11·50 | $c$ = N.E. | 1° 21′ 45″ | sin. = 8·3761729 | | cos. = 9.9998772 |
| | $\delta$ = | 6129·85 | log. = 3·7874499 | | log. = 3·7874499 |
| | D = | 145·755 | log. = 2·1636228 | L = 6128·12 | log. = 3·7873271 |
| 11·50 | $a$ = N.E. | 1° 21′ 45″ | sin. = 8·3761729 | | cos. = 9·9998772 |
| | $\delta$ = | 6147·85 | log. = 3·7887233 | | log. = 3·7887233 |
| | D = | 146·183 | log. = 2·1648962 | L = 6146·11 | log. = 3·7886005 |
| 11·50 | $d$ = N.E. | 1° 26′ 15″ | sin. = 8·3994397 | | cos. = 9·9998633 |
| | $\delta$ = | 6129·85 | log. = 3·7874499 | | log. = 3·7874499 |
| | D = | 153·776 | log. = 2·1868896 | L = 6127·92 | log. = 3·7873132 |

| | | | | | |
|---|---|---|---|---|---|
| 11·50 *b* = N.E. | 1° 26′ 15″ | sin. = 8·3994397 | | cos. = 9·9998633 |
| $\delta$ = | 6147·85 | log. = 3·7887233 | | log. = 3·7887233 |
| D = | 154·288 | log. = 2·1881630 | L = 6145·92 | log. = 3·7885866 |
| 12·00 *c* = N.E. | 1° 49′ 45″ | sin. = 8·5040569 | | cos. = 9·9997786 |
| $\delta$ = | 6129·85 | log. = 3·7874499 | | log. = 3·7874499 |
| D = | 195·662 | log. = 2·2915068 | L = 6126·73 | log. = 3·7872285 |
| 12·00 *a* = N.E. | 1° 41′ 45″ | sin. = 8·5040569 | | cos. = 9·9997786 |
| $\delta$ = | 6147·85 | log. = 3·7887233 | | log. = 3·7887233 |
| D = | 196·237 | log. = 2·2927802 | L = 6144·72 | log. = 3·7885019 |
| 12·00 *d* = N.E. | 1° 54′ 15″ | sin. = 8·5215024 | | cos. = 9·9997601 |
| $\delta$ = | 6129·85 | log. = 3·7874499 | | log. = 3·7874499 |
| D = | 203·682 | log. = 2·3089523 | L = 6126·46 | log. = 3·7872100 |
| 12·00 *b* = N.E. | 1° 54′ 15″ | sin. = 8·5215024 | | cos. = 9·9997601 |
| $\delta$ = | 6147·85 | log. = 3·7887233 | | log. = 3·7887233 |
| D = | 204·280 | log. = 2·3102257 | L = 6144·45 | log. = 3·7884834 |
| 12·50 *c* = N.E. | 2° 17′ 45″ | sin. = 8·6027015 | | cos. = 9·9996513 |
| $\delta$ = | 6129·85 | log. = 3·7874499 | | log. = 3·7874499 |
| D = | 245·557 | log. = 2·3901514 | L = 6124·93 | log. = 3·7871012 |
| 12·50 *a* = N.E. | 2° 17′ 45″ | sin. = 8·6027015 | | cos. = 9·9996513 |
| $\delta$ = | 6147·85 | log. = 3·7887233 | | log. = 3·7887233 |
| D = | 246·177 | log. = 2·3914248 | L = 6142·92 | log. = 3·7883746 |
| 12·50 *d* = N.E. | 2° 22′ 15″ | sin. = 8·6166545 | | cos. = 9·9996281 |
| $\delta$ = | 6129·85 | log. = 3·7874499 | | log. = 3·7874499 |
| D = | 253·574 | log. = 2·4041044 | L = 6124·60 | log. = 3·7870780 |
| 12·50 *b* = N.E. | 2° 22′ 15″ | sin. = 8·6166545 | | cos. = 9·9996281 |
| $\delta$ = | 6147·85 | log. = 3·7887233 | | log. = 3·7887233 |
| D = | 254·318 | log. = 2·4053778 | L = 6142·59 | log. = 3·7883514 |

| | | | | | | |
|---|---|---|---|---|---|---|
| 13·00 $c$ | = N.E. | 2° 45′ 45″ | sin. = 8·6830114 | | | cos. = 9·9995011 |
| $\delta$ | = | 6129·85 | log. = 3·7874499 | | | log. = 3·7874499 |
| D | = | 295·435 | log. = 2·4704613 | L = 6122·81 | | log. = 3·7869510 |
| 13·00 $a$ | = N.E. | 2° 45′ 45″ | sin. = 8·6830114 | | | cos. = 9·9995011 |
| $\delta$ | = | 6147·85 | log. = 3·7887233 | | | log. = 3·7887233 |
| D | = | 296·302 | log. = 2·4717347 | L = 6140·79 | | log. = 3·7882244 |
| 13·50 station | N.E. | 3° 16′ 00″ | sin. = 8·7557469 | | | cos. = 9·9992938 |
| $\delta$ | = | 6138·85 | log. = 3·7880870 | | | log. = 3·7880870 |
| D | = | 349·811 | log. = 2·5438339 | L = 6128·87 | | log. = 3·7873808 |

Having computed all the points connected with the centre C, and carried the results into the preceding table of bearings, distances, etc., we next compute the relative situation of the points A and B from station 10. We have before stated that the line 10 A made an angle with the radius of the curve of 20°; the radius being taken to bear due south from 10, gives the bearing of 10 A = S. 20° W. The angle at A, in the triangle 10 A B, being 83°, gives the bearing of A B = S. 77 E. And the angle at 10, being 65°, gives the bearing 10 to B = S. 45° E. Having thus ascertained the bearings, and the distances being already computed, we now make up a table of bearings and distances, leaving room to put in the latitudes and departures when obtained.

### BEARINGS AND DISTANCES,

### FROM STATION 10 TO A AND B, AND FROM A TO B.

| Bearings and Distances. | Southings. | Eastings. | Westing. |
|---|---|---|---|
| From 10 to B = S.E. 45° = 374·61 feet | 264·884 | 264·884 | |
| " 10 to A = S.W. 20° = 200·00 feet | 187·939 | | 68·404 |
| " A to B = S.E. 77° = 333·288 feet | 76·945 | 338·288 | |

We now proceed to compute the latitudes and departures.

COMPUTATIONS OF THE LATITUDES AND DEPARTURES.

| | | | |
|---|---|---|---|
| S.E. 45° 00′ 00″ | sin. = 9·8494850 | | cos. = 9·8494850 |
| $d$ = 374·61 | log. = 2·5735710 | | log. = 2·5735710 |
| D = 264·884 | log. = 2·4230560 | L = 264·884 | log. = 2·4230560 |
| S.W. 20° 00′ 00″ | sin. = 9·5340517 | | cos. = 9·9729858 |
| $d$ = 200 feet | log. = 2·3010300 | | log. = 2·3010300 |
| D = 68·404 | log. = 1·8350817 | L = 187·939 | log. = 2·2740158 |
| S.E. 77° 00′ 00″ | sin. = 9·9887239 | | cos. = 9·3520880 |
| $d$ = 342·05 | log. = 2·5340960 | | log. = 2·5340960 |
| D = 333·288 | log. = 2·5228199 | L = 76·945 | log. = 1·8861840 |

Having computed our latitudes and departures, and written them in the above table, it now becomes necessary to ascertain their relative position to the point C. We find by our table, computed from the point C, that station 10 has a northing of 6138·853 feet, and easting it has nothing. We see by the above table that A is west of 10 = 68·404 feet. For the purpose of preventing the necessity of an additional column, we shall note this in our table of northings and eastings from point C, as minus eastings, distinguishing it by the negative sign, thus, (— 68·404,) in the easting column. Now A, being south of station 10 187·939 feet, we subtract this sum from the northing of 10, which gives the northing of A. Thus,

| | | | |
|---|---|---|---|
| Station | 10 | northing = | 6138·853 |
| A from | 10 | " = | — 187·939 |
| Leaves the northing of A = | | | 5950·914 feet |

Then, managing in a like manner with station B, we have B east of 10 = 264·884 feet; and, as 10 has neither easting nor westing, we put this into our table as the proper easting.

For the northing we have

| | | | | |
|---|---|---|---|---|
| Station | 10 | | = | 6138·853 |
| and | " | B | = | — 264·884 |
| Leaves the northing of B | | | = | 5873·969 |

Again: for the purpose of proving a portion of our work, we ascertain the relative position of B from A. We have previously found the situation of A to be

| | | | | | |
|---|---|---|---|---|---|
| A | northing | = | 5950·914 | and easting — | 68·404 |
| B (from the preceding table) | " | = | — 76·945 | " | 338·288 |
| B | " | = | 5873·969 | " | 269·884 |

Adding, as indicated by the algebraic signs, gives the northings and eastings of B as above. We now carry these results into the preceding table of northings and eastings, etc., from C; and it will then contain all the points which are needed.

Having completed our table of bearings, distances, northings, and eastings, etc., from the point or centre C, we next compute the bearings from A and B to the several corners of each pier and abutment noted in the aforesaid table, and to station 13·50 of the general location; the bearing to station 10 being already known. Beginning with corner *b*, in the abutment at station 10·50, the formula may be thus enunciated, applying the affix N to the expression representing the station, for northing, and E for easting, we then have

$$A_N \backsim b_N : 1 :: A_E \backsim b_E : \tan. B = \frac{A_E \backsim b_E}{A_N \backsim b_N}$$

wherein B expresses the bearing sought.

## EXAMPLE OF COMPUTATIONS.

| Station | Computation | | Logarithm |
|---|---|---|---|
| **Station 10·50** | A N ∽ *b* N = 5950·914 ∽ 6147·61 | = 196·696 | log. = 2·2937956 |
| | A E ∽ *b* E = − 68·404 ∽ 54·096 | = 122·500 | log. = 2·0881361 |
| | B′ or bearing from A to 10·50 *b* | = N.E. 31° 54′ 51″ * | tan. = 9·7943405 |
| **Station 10·50** | A N ∽ *d* N = 5950·914 ∽ 6129·61 | = 178·696 | log. = 2·2521148 |
| | A E ∽ *d* E = − 68·404 ∽ 53·938 | = 122·342 | log. = 2·0875756 |
| | B′ or bearing from A to 10·50 *d* | = N.E. 34° 23′ 49″ | tan. = 9·8354608 |
| **Station 11·00** | A N ∽ *a* N = 5950·914 ∽ 6147·10 | = 196·186 | log. = 2·2926680 |
| | A E ∽ *a* E = − 68·404 ∽ 96·119 | = 164·523 | log. = 2·2162266 |
| | Bearing from A to 11·00 *a* | = N.E. 39° 59′ 00″ | tan. = 9·9235586 |
| **Station 11·00** | A N ∽ *b* N = 5950·914 ∽ 6146·97 | = 196·056 | log. = 2·2923801 |
| | A E ∽ *b* E = − 68·404 ∽ 104·165 | = 172·569 | log. = 2·2369628 |
| | Bearing from A to 11·00 *b* | = N.E. 41° 21′ 15″ | tan. = 9·9445827 |
| **Station 11·00** | A N ∽ *c* N = 5950·914 ∽ 6129·100 | = 178·186 | log. = 2·2508736 |
| | A E ∽ *c* E = − 68·404 ∽ 95·838 | = 164·242 | log. = 2·2155842 |
| | Bearing from A to 11·00 *c* | = N.E. 42° 40′ 29″ | tan. = 9·9647106 |
| **Station 11·00** | A N ∽ *d* N = 5950·914 ∽ 6128·97 | = 178·056 | log. = 2·2505566 |
| | A E ∽ *d* E = − 68·404 ∽ 103·860 | = 172·264 | log. = 2·2361945 |
| | Bearing from A to 11·00 *d* | = N.E. 44° 03′ 10″ | tan. = 9·9856379 |
| **Station 11·50** | A N ∽ *a* N = 5950·914 ∽ 6146·11 | = 195·196 | log. = 2·2904709 |
| | A E ∽ *a* E = − 68·404 ∽ 146·183 | = 214·587 | log. = 2·3342064 |
| | Bearing from A to 11·50 *a* | = N.E. 47° 52′ 48″ | tan. = 0·0437355 |
| **Station 11·50** | A N ∽ *b* N = 5950·914 ∽ 6145·920 | = 195·006 | log. = 2·2900480 |
| | A E ∽ *b* E = − 68·404 ∽ 154·228 | = 222·632 | log. = 2·3475876 |
| | Bearing from A to 11·50 *b* | = N.E. 48° 47′ 04″ | tan. = 0·0575396 |

* The northing of A being less than the northing of *b*, and the easting of A being less than *b*, the bearing must of course be northeasterly.

| Station | | | |
|---|---|---|---|
| Station 11·50 | AN ∽ *c*N = | 5950·914 ∽ 6128·120 = 177·206 | log. = 2·2484784 |
| | AE ∽ *c*E = — | 68·404 ∽ 145·755 = 214·159 | log. = 2·3307364 |
| | Bearing from A to 11·50 *c* | = N.E. 50° 23′ 38″ | tan. = 0·0822580 |
| Station 11·50 | AN ∽ *d*N = | 5950·914 ∽ 6127·920 = 177·006 | log. = 2·2479880 |
| | AE ∽ *d*E = | — 68·404 ∽ 153·176 = 221·580 | log. = 2·3455306 |
| | Bearing from A to 11·50 *d* | = N.E. 51° 22′ 56″ | tan. = 0·0975426 |
| Station 12·00 | AN ∽ *a*N = | 5950·914 ∽ 6144·720 = 193·806 | log. = 2·2873672 |
| | AE ∽ *a*E = | — 68·404 ∽ 196·237 = 264·641 | log. = 2·4226571 |
| | Bearing from A to 12·00 *a* | = N.E. 53° 47′ 00″ | tan. = 0·1352899 |
| Station 12·00 | AN ∽ *b*N = | 5950·914 ∽ 6144·450 = 193·536 | log. = 2·2867617 |
| | AE ∽ *b*E = | — 68·404 ∽ 204·280 = 272·684 | log. = 2·4356597 |
| | Bearing from A to 12·00 *b* | = N.E. 54° 38′ 06″ | tan. = 0·1488980 |
| Station 12·00 | AN ∽ *c*N = | 5950·914 ∽ 6126·730 = 175·816 | log. = 2·2450584 |
| | AE ∽ *c*E = | — 68·404 ∽ 195·662 = 264·066 | log. = 2·4217125 |
| | Bearing from A to 12·00 *c* | = N.E. 56° 20′ 39″ | tan. = 0·1766541 |
| Station 12·00 | AN ∽ *d*N = | 5950·914 ∽ 6126·460 = 175·546 | log. = 2·2447125 |
| | AE ∽ *d*E = | — 68·404 ∽ 203·682 = 272·086 | log. = 2·4347062 |
| | Bearing from A to 12·00 *d* | = N.E. 57° 09′ 04″ | tan. = 0·1899937 |
| Station 12·50 | AN ∽ *a*N = | 5950·914 ∽ 6142·920 = 192·006 | log. = 2·2833148 |
| | AE ∽ *a*E = | — 68·404 ∽ 246·177 = 314·581 | log. = 2·4977325 |
| | Bearing from A to 12·50 *a* | = N.E. 58° 35′ 37″ | tan. = 0·2144177 |
| Station 12·50 | AN ∽ *b*N = | 5950·914 ∽ 6142·590 = 191·676 | log. = 2·2825677 |
| | AE ∽ *b*E = | — 68·404 ∽ 254·318 = 322·722 | log. = 2·5088286 |
| | Bearing from A to 12·50 *b* | = N.E. 59° 17′ 33″ | tan. = 0·2262609 |
| Station 12·50 | AN ∽ *c*N = | 5950·914 ∽ 6124·930 = 174·016 | log. = 2·2405892 |
| | AE ∽ *c*E = | — 68·404 ∽ 245·577 = 313·961 | log. = 2·4968757 |
| | Bearing from A to 12·50 *c* | = N.E. 61° 00′ 08″ | tan. = 0·2562865 |

| | | | |
|---|---|---|---|
| Station 12·50 | $A_N \backsim d_N$ = 5950·914 ∽ 6124·600 = 173·686 | log. = 2·2397648 |
| | $A_E \backsim d_E$ = − 68·404 ∽ 253·574 = 325·978 | log. = 2·5078262 |
| | Bearing from A to 12·50 $d$ = N.E. 61° 39′ 22″ | tan. = 0·2680614 |
| Station 13·00 | $A_N \backsim a_N$ = 5950·914 ∽ 6140·790 = 189·876 | log. = 2·2784701 |
| | $A_E \backsim a_E$ = − 68·404 ∽ 296·302 = 364·706 | log. = 2·5619430 |
| | Bearing from A to 13·00 $a$ = N.E. 62° 29′ 50″ | tan. = 0·2834729 |
| Station 13·00 | $A_N \backsim c_N$ = 5950·914 ∽ 6122·810 = 171·896 | log. = 2·2352658 |
| | $A_E \backsim c_E$ = − 68·404 ∽ 295·435 = 363·839 | log. = 2·5609065 |
| | Bearing from A to 13·00 $c$ = N.E. 64° 42′ 41″ | tan. = 0·3256407 |
| Station 13·50 | $A_N \backsim 13{\cdot}50_N$ = 5950·914 ∽ 6128·870 = 177·956 | log. = 2·2502980 |
| | $A_E \backsim 13{\cdot}50_E$ = − 68·404 ∽ 349·821 = 418·215 | log. = 2·6213996 |
| | Bearing from A to station 13·50 = N.E. 66° 57′ 01″ | tan. = 0·3711016 |

Having computed the bearings from A to the corners of each pier and abutment shown in the diagram, including stations 10 and 13·50 in the alinement of the road, we now proceed to compute the bearing of the corners of each pier, and the stations 10 and 13·50 from station B.

| | | |
|---|---|---|
| Station 10·50 | $B_N \backsim b_N$ = 5873·969 ∽ 6147·610 = 273·641 | log. = 2·4371812 |
| | $B_E \backsim b_E$ = 264·884 ∽ 54·096 = 210·788 | log. = 2·3238459 |
| | Bearing from B to 10·50 B = N.W. 37° 36′ 27″ | tan. = 9·8866647 |
| Station 10·50 | $B_N \backsim d_N$ = 5873·969 ∽ 6129·610 = 255·641 | log. = 2·4076305 |
| | $B_E \backsim d_E$ = 264·884 ∽ 53·938 = 210·946 | log. = 2·3241713 |
| | Bearing from B to 10·50 $d$ = N.W. 39° 31′ 41″ | tan. = 9·9165408 |
| Station 11·00 | $B_N \backsim a_N$ = 5873·969 ∽ 6147·100 = 273·131 | log. = 2·4363710 |
| | $B_E \backsim a_E$ = 264·884 ∽ 96·119 = 168·765 | log. = 2·2272824 |
| | Bearing from B to 11·00 $a$ = N.W. 31° 42′ 41″ | tan. = 9·7909114 |
| Station 11·00 | $B_N \backsim b_N$ = 5873·969 ∽ 6146·970 = 273·001 | log. = 2·4361642 |
| | $B_E \backsim b_E$ = 264·884 ∽ 104·165 = 160·719 | log. = 2·2060672 |
| | Bearing from B to 11·00 $b$ = N.W. 30° 29′ 09″ | tan. = 9·7699030 |

| | | | |
|---|---|---|---|
| Station 11·00 | $B_N \backsim c_N$ = | 5873·969 ∽ 6129·106 = 255·131 | log. = 2·4067632 |
| | $B_E \backsim c_E$ = | 264·884 ∽ 95·838 = 169·046 | log. = 2·2280049 |
| | Bearing from B to 11·00 *b* | = N.W. 33° 31′ 40″ | tan. = 9·8212417 |
| Station 11·00 | $B_N \backsim d_N$ = | 5873·969 ∽ 6128·970 = 255·001 | log. = 2·4065419 |
| | $B_E \backsim d_E$ = | 264·884 ∽ 103·860 = 161·024 | log. = 2·2068906 |
| | Bearing from B to 11·00 *d* | = N.W. 32° 16′ 15″ | tan. = 9·8003487 |
| Station 11·50 | $B_N \backsim a_N$ = | 5873·969 ∽ 6146·110 = 272·141 | log. = 2·4347940 |
| | $B_E \backsim a_E$ = | 264·884 ∽ 146·183 = 118·701 | log. = 2·0744544 |
| | Bearing from B to 11·50 *a* | = N.W. 23° 33′ 56″ | tan. = 9·6396604 |
| Station 11·50 | $B_N \backsim b_N$ = | 5873·969 ∽ 6145·920 = 271·951 | log. = 2·4344907 |
| | $B_E \backsim b_E$ = | 264·884 ∽ 154·228 = 110·656 | log. = 2·0439751 |
| | Bearing from B to 11·50 *b* | = N.W. 22° 08′ 29″ | tan. = 9·6094844 |
| Station 11·50 | $B_N \backsim c_N$ = | 5873·969 ∽ 6128·120 = 254·151 | log. = 2·4050918 |
| | $B_E \backsim c_E$ = | 264·884 ∽ 145·755 = 119·129 | log. = 2·0760175 |
| | Bearing from B to 11·50 *c* | = N.W. 25° 06′ 51″ | tan. = 9·6709257 |
| Station 11·50 | $B_N \backsim d_N$ = | 5873·969 ∽ 6127·920 = 253·951 | log. = 2·4047499 |
| | $B_E \backsim d_E$ = | 264·884 ∽ 153·176 = 111·708 | log. = 2·0480883 |
| | Bearing from B to 11·50 *d* | = N.W. 23° 44′ 38″ | tan. = 9·6433384 |
| Station 12·00 | $B_N \backsim a_N$ = | 5783·969 ∽ 6144·720 = 270·751 | log. = 2·4325701 |
| | $B_E \backsim a_E$ = | 264·884 ∽ 196·237 = 68·647 | log. = 1·8366216 |
| | Bearing from B to 12·00 *a* | = N.W. 14° 13′ 38″ | tan. = 9·4040515 |
| Station 12·00 | $B_N \backsim b_N$ = | 5783·969 ∽ 6144·450 = 270·481 | log. = 2·4321368 |
| | $B_E \backsim b_E$ = | 264·884 ∽ 204·280 = 60·604 | log. = 1·7825013 |
| | Bearing from B to 12·00 *b* | = N.W. 12° 37′ 45″ | tan. = 9·3503645 |
| Station 12·00 | $B_N \backsim c_N$ = | 5873·969 ∽ 6126·730 = 252·761 | log. = 2·4027100 |
| | $B_E \backsim c_E$ = | 264·884 ∽ 195·662 = 69·222 | log. = 1·8402441 |
| | Bearing from B to 12·00 *c* | = N.W. 15° 18′ 56″ | tan. = 9·4375341 |

| | | |
|---|---|---|
| Station 12·00 | $B_N \backsim d_N = 5873·969 \backsim 6126·460 = 252·491$ | log. = 2·4022459 |
| | $B_E \backsim d_E = 264·884 \backsim 203·682 = 61·202$ | log. = 1·7867656 |
| | Bearing from B to 12·00 *d* = N.W. 13° 37′ 31″ | tan. = 9·3845197 |
| Station 12·50 | $B_N \backsim a_N = 5873·969 \backsim 6142·920 = 268·951$ | log. = 2·4296731 |
| | $B_E \backsim a_E = 264·884 \backsim 246·177 = 18·707$ | log. = 1·2720041 |
| | Bearing from B to 12·50 *a* = N.W. 3° 58′ 44″ | tan. = 8·8423310 |
| Station 12·50 | $B_N \backsim b_N = 5873·969 \backsim 6142·590 = 268·621$ | log. = 2·4291399 |
| | $B_E \backsim b_E = 264·884 \backsim 254·318 = 10·566$ | log. = 1·0239106 |
| | Bearing from B to 12·50 *b* = N.W. 2° 15′ 09″ | tan. = 8·5947707 |
| Station 12·50 | $B_N \backsim c_N = 5873·969 \backsim 6124·930 = 250·961$ | log. = 2·3996062 |
| | $B_E \backsim c_E = 264·884 \backsim 245·557 = 19·327$ | log. = 1·2861644 |
| | Bearing from B to 12·50 *c* = N.W. 4° 24′ 13″ | tan. = 8·8865582 |
| Station 12·50 | $B_N \backsim d_N = 5873·969 \backsim 6124·600 = 250·631$ | log. = 2·3990348 |
| | $B_E \backsim d_E = 264·884 \backsim 253·574 = 11·310$ | log. = 1·0534626 |
| | Bearing from B to 12·50 *d* = N.W. 2° 35′ 02″ | tan. = 8·6544278 |
| Station 13·00 | $B_N \backsim a_N = 5873·969 \backsim 6140·790 = 266·821$ | log. = 2·4262200 |
| | $B_E \backsim a_E = 264·884 \backsim 296·302 = 31·418$ | log. = 1·4971785 |
| | Bearing from B to 13·00 *a* = N.E. 6° 42′ 56″ | tan. = 9·0709585 |
| Station 13·00 | $B_N \backsim c_N = 5873·969 \backsim 6122·810 = 248·841$ | log. = 2·3959220 |
| | $B_E \backsim c_E = 264·884 \backsim 295·435 = 30·551$ | log. = 1·4844564 |
| | Bearing from B to 13·00 *c* = N.E. 6° 59′ 25″ | tan. = 9·0885344 |
| Station 13·50 | $B_N \backsim 13·50_N = 5873·969 \backsim 6128·870 = 254·901$ | log. = 2·4063715 |
| | $B_E \backsim 13·50_E = 264·884 \backsim 349·811 = 84·927$ | log. = 1·9290458 |
| | Bearing from B to station 13·50 = N.E. 18° 25′ 37″ | tan. = 9·5226743 |

We have thus completed our computations of bearings, from the points A and B to the corners of the piers, etc. We now arrange them in the following table, and from the bearings between A and

B, and between A and the corner of the piers, etc., we readily deduce the angles required to be measured at A, and in like manner those required to be measured at B.

## A TABLE OF BEARINGS,

### FROM STATIONS A AND B TO THE CORNERS OF THE SEVERAL PIERS AND ABUTMENTS, AND TO STATIONS 10 AND 13·50;

#### WITH THE ANGLES TO MEASURE FROM A AND B TO EACH CORNER AND STATION INDICATED.

| Piers and Corners. | | Bearings from Station A to the points indicated in column 1. | | Bearings from Station B to the points indicated in Column 1. | | Angles at Station A with B, and the points indicated in Column 1. | Angles at Station B with A, and the points indicated in Column 1. |
|---|---|---|---|---|---|---|---|
| | | | ° ′ ″ | | ° ′ ″ | ° ′ ″ | ° ′ ″ |
| 10·00 station | | N.E. | 20 00 00 | N.W. | 45 00 00 | 83 00 00 | 32 00 00 |
| 10·50 | *b* | " | 31 54 51 | " | 37 36 27 | 71 05 09 | 39 23 23 |
| " | *d* | " | 34 23 49 | " | 39 31 41 | 68 36 11 | 37 28 19 |
| 11·00 | *a* | " | 39 59 00 | " | 31 42 41 | 63 01 00 | 45 18 19 |
| " | *b* | " | 41 21 15 | " | 30 29 09 | 61 38 45 | 46 30 51 |
| " | *c* | " | 42 40 29 | " | 33 31 40 | 60 19 31 | 43 28 20 |
| " | *d* | " | 44 03 10 | " | 32 16 15 | 58 56 50 | 44 43 45 |
| 11·50 | *a* | " | 47 52 48 | " | 23 33 56 | 55 07 12 | 53 26 04 |
| " | *b* | " | 48 47 04 | " | 22 08 29 | 54 12 56 | 54 51 31 |
| " | *c* | " | 50 23 38 | " | 25 06 51 | 52 36 22 | 51 53 09 |
| " | *d* | " | 51 22 56 | " | 23 44 38 | 51 37 04 | 53 15 22 |
| 12·00 | *a* | " | 53 47 00 | " | 14 13 38 | 49 13 00 | 62 46 22 |
| " | *b* | " | 54 38 06 | " | 12 37 45 | 48 21 54 | 64 22 15 |
| " | *c* | " | 56 20 39 | " | 15 18 56 | 46 39 21 | 61 41 04 |
| " | *d* | " | 57 09 04 | " | 13 37 31 | 45 50 56 | 63 22 29 |
| 12·50 | *a* | " | 58 35 37 | " | 3 58 44 | 44 24 23 | 73 01 16 |
| " | *b* | " | 59 17 33 | " | 2 15 09 | 43 42 27 | 74 44 51 |
| " | *c* | " | 61 00 08 | " | 4 24 13 | 41 59 52 | 72 35 47 |
| " | *d* | " | 61 39 22 | " | 2 35 02 | 41 20 38 | 74 24 58 |
| 13·00 | *a* | " | 62 29 50 | N.E. | 6 42 56 | 40 30 10 | 83 42 56 |
| " | *c* | " | 64 42 41 | " | 6 59 25 | 38 17 19 | 83 59 25 |
| 13·50 station | | " | 66 57 01 | " | 18 25 37 | 36 02 59 | 95 25 37 |
| | A | | | N.W. | 77 00 00 | | |
| | B | S.E. | 77 00 00 | | | | |

We have thus completed our table of angles which are to be used in the location of the points indicated as follows. Having two observers with instruments, one at station A and the other at B, each having a suitable instrument, they proceed to lay off upon their respective instruments the angles indicated in the table to any one of the corners or stations desired. Having done this, an assistant repairs with a boat to the place of intersection of the lines corresponding with their instruments; and, if the water be not too deep, fixes a stake by driving it into the mud or sand which forms the bottom of the river. Or, if piles are being driven from a scow, the position of the pile may be brought to the intersection indicated by the instruments, and driven. Or, the point may be otherwise marked by mooring a buoy or float by the aid of two or three lines; and doubtless, by many other devices, marks may be fixed which will be found equally simple and exact, the whole of the mechanical operations being so simple in their character as not to need further description. I will only add that we have made use of the method which we have here endeavored to develope in several instances, and have found it very convenient and accurate.

**(51)** We have thus, in the foregoing pages, completed our contemplated essays upon railroad curves connected with the alinements of the main tracks, side tracks, and turnouts; we now propose to add a formula for uniting the different gradients of railroad tracks with vertical curves.

Before we proceed to the investigation of formula, we would remark, it is not our purpose to give anything like a full description of the operations for laying down the gradients of a railroad track, (the operations being so simple in character as to be readily

[Fig. 21.]

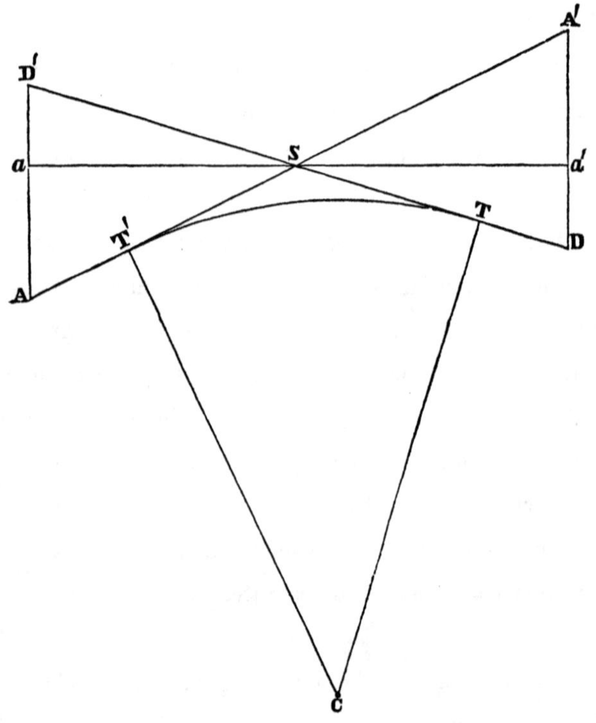

comprehended by every one,) but merely to develope a formula which has been found convenient and useful in our practice of rounding off the salient angles and hollowing the re-entering angles (if they may be so termed) formed by the intersections of the gradients of a railroad track.

(**52**) In our practice we have never laid down a vertical curve of a less radius than forty thousand feet, but in general our curves have embraced two hundred feet upon either side of the intersecting point of the gradients; that is to say, the vertical arc has usually been about 400 feet in length; but, when the inclinations of the gradients have been such as to make the angles to be rounded or hollowed, comparatively acute, we have sometimes used a shorter arc.

Presuming that the inclinations of the gradients and the relative positions of the angles have been determined, we commence with the investigation of formula for determining the value of these angles in degrees.

The problem presents four different cases, viz., the salient angle formed by an ascending and descending grade. The re-entering angles formed by two descending or ascending grades, one of which being much more inclined than the other. The re-entering angle formed by a level line, and one ascending or descending grade. The re-entering angle formed by a descending and an ascending grade.

To explain the foregoing angles, and the method of ascertaining their value in degrees and minutes, (See Fig. 21,) let $a\ a'$ represent a level line, A S an ascending grade, and S D a descending grade, then will the angle at S be a salient angle. To determine the measure of this angle, viz., A S D, we will suppose A S to ascend at the rate of forty feet to the mile, and S D to descend at the rate of twenty-five feet to the mile. We will now suppose S $a$ to equal one mile, or 5280 feet; S D and S A′ are also taken as a mile each; inasmuch as there will be no practical difference between the length of a line inclining 40 or even 80 feet to the mile, and the same line reduced to a level, (and this remark will apply almost universally, or to the gradients of railroads in general;) therefore the lines S A, S $a$, S D′, are taken each as one mile.

From an inspection of the figure, it will be obvious that the angle A S D will equal 180 — (A S $a$ + D S $a'$) and the angle A S $a$ = A′ S $a'$; wherefore, A S $a$ + D S $a'$ = D S A′; and the following method of determining which, though not strictly accurate, will be found sufficiently exact for every practical purpose. Taking S D and S A′ = one mile each, then will $a'$ A′ = 40 feet, and D $a'$ = 25 feet, and the angles S D A′ and S A′ D being each so near a right angle that we may take either of them as such. Taking the angle D as a right angle, S D will be a cosine, and D A′ a sine, and S A′ will be a radius.

Then, taking radius = unity, and representing it by R, we have this analogy, cos. : sin. :: R : tan. = $\frac{\text{sin.}}{\text{cos.}}$ ; which, in practice, will stand thus,

$$\text{S D} : (\text{D } a' + a' \text{ A}') :: \text{R} : \text{tan. A}' \text{ S D} = \frac{(\text{D } a' + a' \text{ A}')}{\text{S D}} ;$$

$$\text{and } 180° - \text{A}' \text{ S D} = \text{A S D};$$

and consequently, the angle A′ S D = C in the quadrilateral C T′ S T. (63)

Having obtained the angles S and C in the quadrilateral, we will now proceed to give an example of computation. As before stated, (D $a'$ + $a'$ A) = D A′ = 25 + 40 = 65 feet, and S D = 5280 feet = 1 mile; then, we have

| | | | |
|---|---|---|---|
| D A′ = | 65 feet | log. = | 1·8129134 |
| S D = | 5280 " | log. = | 3·7226339 |
| C = | 0° 42′ 19″ | tan. = | 8·0902795 |

We take the angle C to the nearest second; it is not necessary that we should be more exact.

Then, taking the distances S T and S T′ = 260 feet each, we ascertain the radius, (in our computations it is not necessary to know the radius, and we merely ascertain it this time as a matter of curiosity rather than use;) to find which, we have

Sin. $\frac{1}{2}$ C : 260 feet :: cos. $\frac{1}{2}$ C : radius = cot. $\frac{1}{2}$ C 260 (64)

| | | | |
|---|---|---|---|
| Thus, | 260 | log. = | 2·4149733 |
| | $\frac{1}{2}$ C = 0° 21′ 10″ | cot. = | 2·2106159 |
| | Radius = 42226 feet | log. = | 4·6255892 |

We have thus found our radius = 42226 feet, the angle S being very acute, (speaking comparatively,) it becomes necessary to take the distances S T and S T′ somewhat greater than it is our habit. We would remark here, that whatever distance we assume for S T, it will be found convenient that it should divide even by the number 20, because in setting the grade-pins for laying down the rails, it is usual to place them 20 feet apart, which is as long as we can conveniently have the straight-edged board used as a guide in placing the sleepers or ties to the proper height or grade.

We have taken S T = 260 feet, which, divided by 20 feet, will

[Fig. 22.]

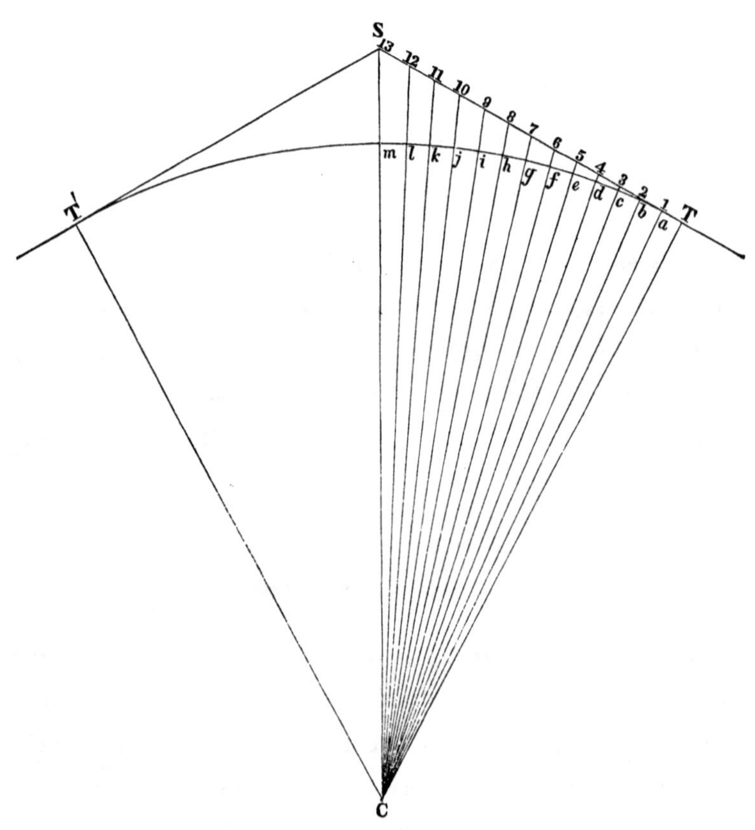

give thirteen divisions upon each side of S; consequently, the portion of the arc spanned by one of these divisions will be equal to $\frac{\frac{1}{2} \text{C}}{13}$.

We have demonstrated in section (2) that the angle of deflection is equal to half the centre angle, spanned by the chord governing said deflection. We therefore have in the large triangle T C 1, the angle at T $= 90$, consequently the angle at 1 is equal to the complement of the angle C; and the angle C $= \frac{\frac{1}{2}\text{ T C T}'}{13} = \frac{0^{\circ}\ 21'\ 10''}{13}$ $= 0^{\circ}\ 01'\ 37''\cdot 7$; hence the angle at 1 $= 89^{\circ}\ 58'\ 22''\cdot 30$, which is common to the triangles T 1 C and T 1 *a;* then, in the triangle T 1 *a*, we have T $= \frac{0^{\circ}\ 01'\ 37''\cdot 70}{2} = 0^{\circ}\ 00'\ 48''\cdot 85$; hence the angle at *a* will be equal to $180^{\circ} - (89^{\circ}\ 58'\ 22''\cdot 30 + 00'\ 48''\cdot 85) =$ $90^{\circ}\ 00'\ 48''\cdot 85$. That is, the angle at *a* is equal to $90^{\circ} +$ the angle at T; and the angle *b*, in the triangle T *b* 2, will also be equal to $90^{\circ} +$ the angle at T; and the same remark will apply to the triangles T *c* 3, T *d* 4, etc., to T *m* S.

Having thus explained the method of deducing the angles, we find the ordinates *a* 1, *b* 2, *c* 3, etc., to *m* S, by the following formula. Taking the triangle T *a* 1 for an example, and making use of the symbols belonging to the same, we have sin. $90^{\circ} +$ T : T 1 :: sin. T : *a* 1; but, since the sin. of $90^{\circ} +$ T is equal to the cos. T, the formula may be rendered thus, cos. T : T 1 :: sin. T : *a* 1 $=$ tan. T $\times$ T 1 (65)

We wish to remark, that in the investigation above, we have considered the arc T′ *m* T and the tangent lines T S and S T of equal length; consequently, the computed ordinates will be practically the same as if they were perpendiculars to the tangent line T S.

We will now show, by actual computations, that the formula, though not in the strictest sense exact, still it presents by far a greater degree of accuracy than would be possible to practise in laying down railroad tracks.

As a test to our formula, we will now determine the length of the curve T′ *m* T, and compare the same with the tangent lines T S, S T′. In the triangle S T C, we have

$$\text{Sin. C} : \text{S T} :: \text{cos. C} : \text{rad.} = \text{cot. C} \times \text{S T}.$$

Then, representing S T by $t$, the radius in seconds by $r''$, and the angle C in seconds by C″, and the radius in the unit of measure (found above) by $r$, and the length of the curve by $c$, we shall have

$$r'' : r :: C'' : c = \frac{r \,.\, C''}{r''} = \frac{\text{cot. C} \,.\, t \,.\, C''}{r''} \qquad (66)$$

| | | | | |
|---|---|---|---|---|
| Thus, | C | = 0° 21′ 00″ | cot. | = 2·2106159240 |
| | $t$ | = 260 feet | log. | = 2·4149733480 |
| | C″ | = 2540 " | log. | = 3·4048337166 |
| | $r''$ | = | co. ar. log. | = 4·6855748668 |
| | $c$ | = 519·9934 feet | log. | = 2·7159978554 |
| | $2t$ | = 520 | | |
| Difference | | = 0·0066 | | |

We find thus, that the arc and the tangent lines agree within $\frac{66}{10000}$ of a foot, which is a little larger than $\frac{1}{16}$ of an inch, a quantity quite too small to be considered an error in laying down a railroad track, especially in the ordinates where the error in the greatest will be reduced in the proportion the length of radius bears to the ordinate, which can never amount to anything worth noticing, especially when we consider that the case we are examining is of that class which produces errors greater in amount than the most of cases which come under consideration; so that I think we may be warranted in pronouncing our formula practically exact.

Having investigated the necessary formula, we will now proceed to give a specimen of calculation.

We have found in the foregoing, the angle T in the triangle T *a* 1 = 00′ 48″·85, and in the triangle T *b* 2 it will be twice that amount, and in the triangle T *c* 3 it will be three times that amount; and so on to the centre ordinate *m* S, which will be thirteen times the amount.

| | | | | | |
|---|---|---|---|---|---|
| Commencing with T *a* 1 | we have | T | = 0° 00′ 48″·85 | tan. = 6·3744395 |
| | | T 1 | = 20 feet | log. = 1·3010300 |
| | | *a* 1 | = 0·00473663 feet | log. = 7·6754695 |
| In the triangle T *b* 2 | we have | T | = 0° 01′ 37″·07 | tan. = 6·6754695 |
| | | T 2 | = 40 feet | log. = 1·6020600 |
| | | *b* 2 | = 0·0189465 feet | log. = 8·2775295 |
| In the triangle T *c* 3 | we have | T | = 0° 02′ 26″·55 | tan. = 6·8515608 |
| | | T 3 | = 60 feet | log. = 1·7781513 |
| | | *c* 3 | = 0·0426297 feet | log. = 8·6297121 |
| In the triangle T *d* 4 | we have | T | = 0° 03′ 15″·40 | tan. = 6·9764996 |
| | | T 4 | = 80 feet | log. = 1·9030900 |
| | | *d* 4 | = 0·0757861 feet | log. = 8·8795896 |
| In the triangle T *e* 5 | we have | T | = 0° 04′ 04″·25 | tan. = 7·0734097 |
| | | T 5 | = 100 feet | log. = 2·0000000 |
| | | *e* 5 | = 0·118415 feet | log. = 9·0734097 |
| In the triangle T *f* 6 | we have | T | = 0° 04′ 53″·01 | tan. = 7·1525910 |
| | | T 6 | = 120 feet | log. = 2·0791812 |
| | | *f* 6 | = 0·170518 feet | log. = 9·2317722 |
| In the triangle T *g* 7 | we have | T | = 0° 05′ 41″·95 | tan. = 7·2195379 |
| | | T 7 | = 140 feet | log. = 2·1461280 |
| | | *g* 7 | = 0·232095 feet | log. = 9·3656659 |

| | | | |
|---|---|---|---|
| In the triangle T $h$ 8 we have | T | = 0° 06′ 30″·08 | tan. = 7·2775300 |
| | T 8 | = 160 feet | log. = 2·2041200 |
| | $h$ 8 | = 0·303144 feet | log. = 9·4816500 |
| In the triangle T $i$ 9 we have | T | = 0° 07′ 19″·65 | tan. = 7·3286827 |
| | T 9 | = 180 feet | log. = 2·2552725 |
| | $i$ 9 | = 0·383668 feet | log. = 9·5839552 |
| In the triangle T $j$ 10 we have | T | = 0° 08′ 08″·50 | tan. = 7·3744403 |
| | T 10 | = 200 feet | log. = 2·3010300 |
| | $j$ 10 | = 0·473664 feet | log. = 9·6754703 |
| In the triangle T $k$ 11 we have | T | = 0° 08′ 57″·35 | tan. = 7·4158332 |
| | T 11 | = 220 feet | log. = 2·3424227 |
| | $k$ 11 | = 0·573134 feet | log. = 9·7582559 |
| In the triangle T $l$ 12 we have | T | = 0° 09′ 46″·20 | tan. = 7·4536218 |
| | T 12 | = 240 feet | log. = 2·3802112 |
| | $l$ 12 | = 0·682076 feet | log. = 9·8338330 |
| In the triangle T $m$ S we have | T | = 0° 10′ 35″·05 | tan. = 7·4883842 |
| | T S | = 260 feet | log. = 2·4149733 |
| | $m$ S | = 0·800493 feet | log. = 9·9033575 |

We have thus computed the thirteen ordinates according to the formula. It will be seen that the first angle in the triangle T $a$ 1 is taken a very small amount too large, which will make all the angles used something large, but not sufficiently so as to practically affect our results.

The reason why we did not correct the angles in the course of our operations was, that we may be enabled to compare the results obtained by another method, which will much abridge the work; and, although not strictly accurate, still we may state, as we have before, respecting the preceding formula, that it is practically exact.

The other method may be explained thus. Having divided the line T S, in a suitable number of parts, which, for the purpose of comparing with our previous computations, we will suppose to be thirteen, of 20 feet each; we must then compute the first ordinate, viz., *a* 1, in the triangle T *a* 1, which we call *y;* which, however, in the present case will be needless, as we have it already computed. We therefore take the value of *y* from our previous computations, which of course will need no comparison; we then find the remainder of the ordinates, *b* 2, *c* 3, *d* 4, etc., to *m* S, according to the expressions given in the following table.

| Column of Ordinates. No. 1. | | Expression of Formulæ. No. 2. | | Computed Results. No. 3. | Computed Results by Previous Formulæ, for comparison. No. 4. |
|---|---|---|---|---|---|
| No. 1 or *a* 1 | = | $y$ | = | 0·00473663 feet | 0·00473663 feet |
| 2 " *b* 2 | = | $2^2 y$ | = | 0·01894652 " | 0·01894650 " |
| 3 " *c* 3 | = | $3^2 y$ | = | 0·04262967 " | 0·0426297 " |
| 4 " *d* 4 | = | $4^2 y$ | = | 0·07578608 " | 0·0757861 " |
| 5 " *e* 5 | = | $5^2 y$ | = | 0·11841575 " | 0·1184150 " |
| 6 " *f* 6 | = | $6^2 y$ | = | 0·17051868 " | 0·1705180 " |
| 7 " *g* 7 | = | $7^2 y$ | = | 0·23209487 " | 0·2320950 " |
| 8 " *h* 8 | = | $8^2 y$ | = | 0·30314432 " | 0·303145 " |
| 9 " *i* 9 | = | $9^2 y$ | = | 0·38366703 " | 0·383668 " |
| 10 " *j* 10 | = | $10^2 y$ | = | 0·47366300 " | 0·473664 " |
| 11 " *k* 11 | = | $11^2 y$ | = | 0·57313223 " | 0·573134 " |
| 12 " *l* 12 | = | $12^2 y$ | = | 0·68207472 " | 0·682076 " |
| 13 " *m* S | = | $13^2 y$ | = | 0·80049047 " | 0·800493 " (67) |

Explanation of the Tables. The first column contains the number of the ordinates, arranged in numerical order, from one to thirteen. The second column contains the notation expressing the method of computation. Third column contains the resulting computations. Fourth column contains the same elements, computed by the preceding formula, which is placed here for the purpose of conveniently comparing the results of the two methods.

The comparison shows, that the method by squares, is sufficiently accurate for the most exact work, when we consider we can only make use of the three first decimals, in practice, while the two methods do not differ at all in the fifth decimal. I repeat, we may without hesitation, pronounce the rule practically exact.

**(53)** In applying the foregoing results to practice, doubtless different engineers will pursue different methods; but, a convenient method is to ascertain the total heights (as they are generally called) of stations in the inclined lines, which shall correspond to stations of the same number, belonging to the vertical curves; then adding or subtracting the computed elements of the curve, or length of ordinates, corresponding to the station set out in the inclined lines, accordingly as the nature of the case requires.

**(54)** It sometimes happens that the locating stations are obliterated, and in that case, the position of the angle at S, (which is the main starting point,) cannot be readily found. In such cases the engineer must measure a portion of the road anew, extending the measurement sufficiently far from the apex or point S, upon both sides, so as to ascertain the grades correctly. Having completed the measurements, and marked and numbered the stations, (which are usually fixed one hundred feet apart,) and determined their relative levels, and the inclination of the grades which govern our operations, we proceed to ascertain the points of intersection.

First, draw the line $k\ k'$ to represent a level, (See Fig. 23 ($a$);) then, draw the line $n\ l'$, at an angle with $k\ k'$, equal to the inclination of the grade S $n$. Again, through the point S, draw the line $n'\ l$ at an angle with $k\ k'$, equal to the inclination of the grade S $n'$. From the point $n$ let fall the perpendicular $n\ m$; and also, from the point $n'$, let fall the perpendicular $n'\ l'$, until it intersects the line $n\ l'$. Then, from the intersection of $n\ l'$ with $n'\ l'$, draw the line $l'\ m$ parallel with $k\ k'$; and, from the intersection of $n\ l$ with $n'\ l$, draw the line $l\ m'$ parallel with $k\ k'$.

It will now be obvious, from an inspection of the diagram, that the angle $n$ S $k$ is equal to the grade or inclination of $n$ S; and the angle $k$ S $l$ is equal to the grade or inclination of S $n'$. And it will also be obvious that the angle $n'$ S $k'$ will be equal to the grade or inclination S $n'$; and the angle $k'$ S $l'$ will be equal to the grade or inclination S $n$. Then, representing by

$d$, the distance between the stations, (usually 100 feet;)

$g$, the difference in heights between the stations in grade $n$ S;

$g'$, " " " " " $n'$ S;

$n$ and $n'$, the numbers of the stations at the points they represent;

$h$, the height of station $n$;

$h'$, the " " $n'$; we have

$$d : g :: n \backsim n' : n\,m;$$

$$\text{and } d : g' :: n \backsim n' : n'\,m' \qquad (68)$$

Then, by substituting $p$ for $n\ m$, and $p'$ for $n'\ m'$; $\delta$ for $n$ S, and $\delta'$ for $n'$ S; and, subtracting $p$ from $h$, (which gives the height of the line $m\ l'$ above the datum line, and which we will represent by $o'$;) then, subtracting $p'$ from $h'$, (which gives the height of $m'\ l$ above the datum line, and which we will represent by $o$,) we have

[Fig. 23 (a).]

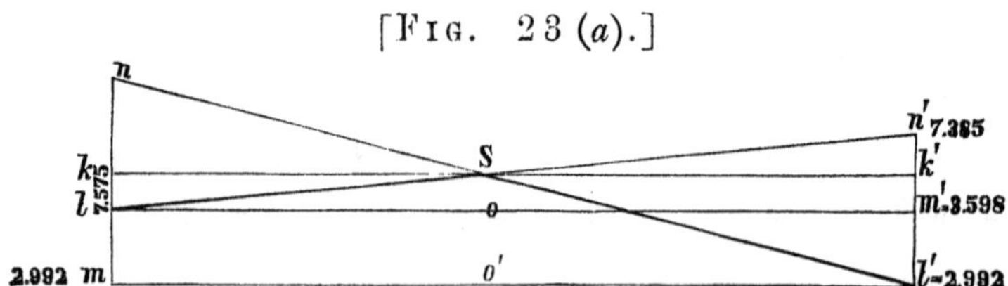


[Fig. 23 (b).]

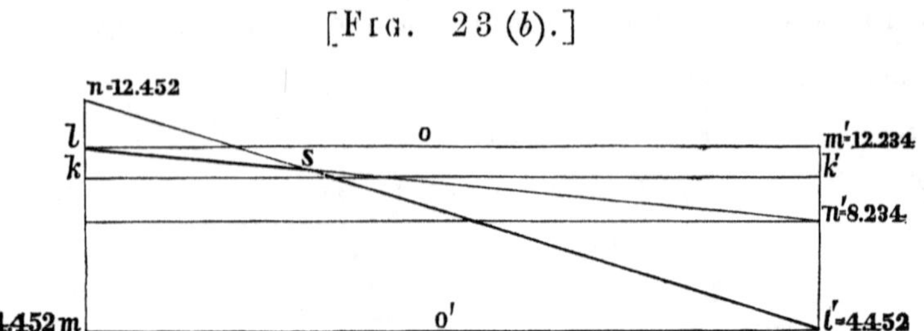

$$g + g' : d :: (h - o) : \delta = \frac{(h - o) \,.\, d}{g + g'}$$

$$g + g' : d :: (h' - o') : \delta' = \frac{(h' - o') \,.\, d}{g + g'} \qquad (69)$$

We will suppose, for an example of calculation, station $n$ to be numbered 10, and station $n'$ to be numbered 20; $h$, to be 10·567 feet above the datum line, and $h'$ to be 7·385 feet above the datum line; $g$ to be 0·7575 feet, and $g'$ to be $=$ 0·3787; and both grades to be ascending from S; with $d = 100$ feet. We then have

| | | | | |
|---|---|---|---|---|
| | $d$ | = 100 feet | co. ar. | log. = 8·0000000 |
| | $g$ | = 0·7575 feet | | log. = 9·8793826 |
| | $n \backsim n'$ | = 10 stations, or 1000 feet | | log. = 3·0000000 |
| | $n\ m = p$ | = 7·575 feet | | log. = 0·8793826 |
| | $h$ | = 10·567 | | |
| | $p - h = o'$ | 2·992 | | |
| Again, | $d$ | = 100 feet | co. ar. | log. = 8·0000000 |
| | $g'$ | = 0·3787 feet | | log. = 9·5782953 |
| | $n \backsim n'$ | = 10 stations, or 1000 feet | | log. = 3·0000000 |
| | $n'\ m' = p'$ | = 3·787 feet | | log. = 0·5782953 |
| | $h'$ | = 7·385 | | |
| | $h' - p = o$ | = 3·598 | | |
| | $g' + g$ | = 1·1362 feet | co. ar. | log. = 9·9445452 |
| | $d$ | = 100 feet | | log. = 2·0000000 |
| | $h - o$ | = 6·969 | | log. = 0·8431705 |
| | $\delta$ | = 6·1336 | | log. = 2·7877157 |
| | $g + g'$ | = 1·1362 | co. ar. | log. = 2·9445452 |
| | $d$ | = 100 feet | | log. = 2·0000000 |
| | $h - o'$ | = 4·395 | | log. = 0·6427612 |
| | $\delta$ | = 3·866 | | log. = 2·5873064 |

Having thus ascertained the distances $n$ S, and $n'$ S $= \delta$ and $\delta'$; if we now take $n + \delta =$ the number of the station, represented by S, (the point of the intersection of the grades;) then will $n - \delta$ $=$ the same number S, if the computations be correctly prepared.

| | | Stations. | | Stations. |
|---|---|---|---|---|
| Thus, | $n$ | $= 10$ | $n'$ | $= 20$ |
| | $\delta$ | $= 6 \cdot 1336$ | $\delta'$ | $= 3 \cdot 8664$ |
| | $n + \delta$ | $= 16 \cdot 1336$ | $n' - \delta'$ | $= 16 \cdot 1336$ |

We may now further prove our work by ascertaining the height of the point S, from the datum line, by computing the descent from $n$ to S, and from $n'$ to S. If our computations are correct, the results should be alike. Thus, we have $h - \delta.g =$ the height of S; and $h' - \delta'.g' =$ height of S.

| | | | |
|---|---|---|---|
| | $\delta$ | $= 6 \cdot 1336$ | log. $= 0 \cdot 7877155$ |
| | $g$ | $= 0 \cdot 7575$ | log. $= 9 \cdot 8793826$ |
| | $\delta . g$ | $= 4 \cdot 6462$ | $0 \cdot 6670981$ |
| | $h$ | $= 10 \cdot 567$ | |
| Height of | S | $= 5 \cdot 9208$ above datum line. | |
| | $\delta'$ | $= 3 \cdot 8664$ | log. $= 0 \cdot 5873068$ |
| | $g'$ | $= 0 \cdot 3787$ | log. $= 9 \cdot 5782953$ |
| | $\delta' . g'$ | $= 1 \cdot 4642$ | $0 \cdot 1656021$ |
| | $h'$ | $= 7 \cdot 385$ | |
| Height of | S | $= 5 \cdot 9208$ above datum line. | |

Having thus found the station corresponding to the intersecting point of the grades, and its relative height, the necessary stations for laying down the vertical curves can be readily prepared, and the work can be proceeded with in the manner set forth in the foregoing.

To make the formula just enunciated applicable to every case would require several modifications; we shall, however, only give one, believing that the ingenuity of the reader will readily supply whatever may be deficient.

**(55)** The case we propose, is, when we have one grade descending say 0·800 feet per 100 feet, which will intersect another descending grade of 0·400 feet per 100 feet.

To describe the construction of a figure applicable to this case we have only to copy verbatim the description of Fig. 23 (*a*); we therefore refer to that description as a substitute.

After having constructed the figure, it will be obvious that we have but few modifications to make in the formula already given; but, lest we should not be fully understood, we repeat our former formula, with the necessary modifications. Thus we have

$$d : g :: n \backsim n' : n\,m; \text{ and } d : g' :: n \backsim n' : n'\,m' \qquad (70)$$

Then, as before, substituting $p$ for $n\,m$, and $p'$ for $n'\,m'$; $\delta$ for $n$ S, and $\delta'$ for $n$ S; and then, subtracting $p$ from $h$, we get the height of $m\,l'$ above the datum line, (which height we represent by $o'$;) and then, by adding $p'$ to $h'$ we obtain the height of $m'\,l$ above the datum line, (which height we represent by $o$.) We then have

$$g \backsim g' : d :: h - o : \delta$$

$$\text{and } g \backsim g' : d :: h' - o' : \delta' \qquad (71)$$

EXAMPLE OF COMPUTATION.

Let $d = 100$ feet; $g = 0{\cdot}800$ feet; $g' = 0{\cdot}400$ feet; and $n = 10$; $n' = 20$; $h = 12{\cdot}452$; $h' = 8{\cdot}234$. We then have

$$\text{Firstly, } d : g :: n \backsim n' : n\,m;$$

$$\text{Secondly, } d : g' :: n \backsim n' : n'\,m'.$$

| Thus, | | | | |
|---|---|---|---|---|
| | $d$ | = 100 feet | co. ar. | log. = 8·0000000 |
| | $g$ | = 0·800 feet | | log. = 9·9030900 |
| | $n \backsim n'$ | = 10 stations, or 1000 feet | | log. = 3·0000000 |
| | $n'\,m = p$ | = 8·00 feet | | log. = 0·9030900 |

[FIG. 24.]

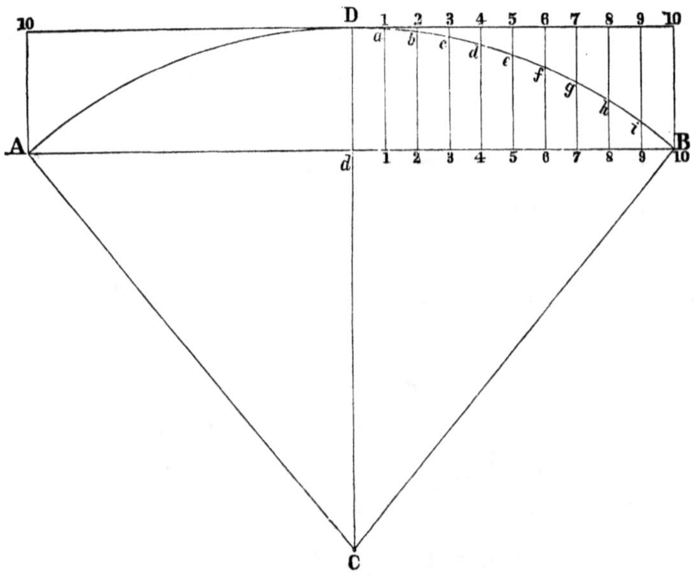

| Secondly, | | | | |
|---|---|---|---|---|
| $d$ | = 100 feet | co. ar. | log. = 8·0000000 |
| $g'$ | = 0·400 feet | | log. = 9·6020600 |
| $n \backsim n'$ | = 10 stations, or 1000 feet | | log. = 3·0000000 |
| $n' m = p'$ | = 4·00 feet | | log. = 0·6020600 |

| | | | |
|---|---|---|---|
| $h$ | = 12·452 | $h'$ | = 8·234 |
| $p$ | = 8·000 | $p'$ | = 4·000 |
| $(p - h) = o' =$ | 4·452 | $(p' + h') = o =$ | 12·234 |
| $h'$ | = 8·234 | $h$ | 12·452 |
| $(h' - o')$ | = 3·782 | $(h - o)$ | = 0·218 |

Again, we have

$$g \backsim g' : d :: h - o : \delta;$$

$$\text{and } g \backsim g' : d :: h' - o' : \delta \qquad (71)$$

| | | | |
|---|---|---|---|
| $g \backsim g'$ = 0·400 | | co. ar. | log. = 0·3979400 |
| $d$ = 100 feet | | | log. = 2·0000000 |
| $h - 0$ = 0·218 feet | | | log. = 9·3384565 |
| $\delta$ = 0·545 stations, or 54·5 feet | | | log. = 1·7363965 |
| $g \backsim g'$ = 0·400 | | co. ar. | log. = 0·3979400 |
| $d$ = 100 feet | | | log. = 2·0000000 |
| $h' - o'$ = 3·782 feet | | | log. = 0·5777215 |
| $\delta'$ = 945·5 feet, or 9·455 stations | | | log. = 2·9756615 |

(**56**) We conclude our remarks upon tracklaying with a formula for the computation of elements convenient for setting out curving boards or patterns for bending rails to suit the horizontal curves of short radii.

The principles of our present formula are based upon a chord of the arc equal in length to the longest rails, being divided into equal spaces or abscissa; and then ascertaining the length of the corresponding ordinates which shall extend therefrom to the periphery or curve. Calculations based upon strict formula, being somewhat lengthy, will require considerable time and labor to perform them. The engineer being frequently called upon to give, in great haste, the elements for making a pattern to guide the tracklayer in curving his rails, it becomes desirable to obtain a formula as short and convenient as practicable.

These considerations have led to the adoption of the following formula, which, though not strictly correct, is nevertheless as accurate as mechanical skill requires.

By way of explanation, suppose it desirable to form a pattern for bending rails of twenty feet in length, it will be found convenient to divide the chord into equal parts of one foot each.

From an examination of the sketch, it will be obvious that one of these divisions will bisect both the chord and the arc, and that the parts thus bisected will be similar and equal; therefore, the computations made for the one part will apply to the other.

To proceed to the investigation, we first ascertain the angle at the centre of the curve spanned by an absciss at the periphery of one foot.

Representing this angle by C′; the absciss by $a$; the ordinate corresponding to No. 1, by $y$; and the radius of the curve by $r$; then, by considering $r$ a cosine; and the absciss $a$, which spans the arc, a sine; we have the following analogy,

$$\text{Cos.} : \text{sin.} :: \text{R} : \text{tan. C}';$$

which, by substituting for the cosine its value $= r$; and, for the sine, its value $= a =$ unity; we then have

Tan. $C = \frac{1}{r}$ and cos. $\frac{1}{2}$ C : $a$ :: sin. $\frac{1}{2}$ C : $y =$ tan. $\frac{1}{2}$ C′ . $a =$ tan. $\frac{1}{2}$ C′ . 1; it is now obvious that $\frac{C}{2} = \frac{\frac{1}{r}}{2} = \frac{1}{r \cdot 2}$; wherefore, tan. $\frac{1}{2}$ C . $1 = y = \frac{1}{r \cdot 2}$ (72)

Performing the computations indicated, by logarithms, we have log. $y =$ (ar. co. log. $r +$ ar. co. log. 2.)

It will be seen by the above expression that we have considered the arc and the tangent of the same length, which will be found sufficiently exact for every practical purpose, and that the ordinate represented by $y = 1\ a$, as represented in the figure. To find the remainder of the ordinates we have, for ordinate

| | | | | |
|---|---|---|---|---|
| No. 2 | = | $b$ 2 | = | $2^2\ y$ |
| 3 | = | $c$ 3 | = | $3^2\ y$ |
| 4 | = | $d$ 4 | = | $4^2\ y$ |
| 5 | = | $e$ 5 | = | $5^2\ y$ |

etc., to the number of ordinates contained in half the chord or tangent line.

For the purpose of testing the degree of accuracy of the formula enunciated above, it may be necessary to obtain from strict computations the ordinate D $d$ = (10) (10.) This ordinate will correspond to $10^2\ y$, or the greatest ordinate of the computation, and will contain a greater error than any one of the others.

Investigation of Exact Formula. Let $r$ represent the radius of the curve; R the radius of the tables; $\frac{1}{2}\ ch$ the half of the chord AB; C the angle D C B; then will $r$ : R :: $\frac{1}{2}\ ch$ : sin. $C = \frac{\frac{1}{2}\ ch}{r}$; and

Cos. $\frac{1}{2}$ C : $\frac{1}{2}\ ch$ :: sin. $\frac{1}{2}$ C : D $d =$ tan. $\frac{1}{2}$ C . $\frac{1}{2}\ ch$ (74)

which expression corresponds to the greatest ordinate of the calculations; and which, in the practical examples we shall give, will be represented by $10^2 y$. So that the difference between the ordinate found by the exact formula and $10^2 y$ will constitute the error.

Example of computation, according to the approximate formula. Assuming $r = 300$ feet; $\frac{1}{2}\ ch = 10$ feet; and $a = 1$ foot; we shall then have

| | | | |
|---|---|---|---|
| $r$ | = 300 feet | co. ar. | log. = 7·5228787 |
| 2 | = | co. ar. | log. = 9·6989700 |
| $y$ | = 0·0016666 feet | | log. = 7·2218487 |

| | | | | |
|---|---|---|---|---|
| Having found | No. 1 = $a$ 1 | = $y$ | = 0·0016666 |
| we have | 2 = $b$ 2 | = $2^2\ y$ | = 0·0066666 |
| | 3 = $c$ 3 | = $3^2\ y$ | = 0·0149999 |
| | 4 = $d$ 4 | = $4^2\ y$ | = 0·0266666 |
| | 5 = $e$ 5 | = $5^2\ y$ | = 0·0416666 |
| | 6 = $f$ 6 | = $6^2\ y$ | = 0·0599999 |
| | 7 = $g$ 7 | = $7^2\ y$ | = 0·0816666 |
| | 8 = $h$ 8 | = $8^2\ y$ | = 0·1066666 |
| | 9 = $i$ 9 | = $9^2\ y$ | = 0·1349999 |
| | 10 = $j$ 10 | = $10^2\ y$ | = 0·1666666 |

Example of computation of the greatest ordinate, according to the exact formula, as a test to the above, viz., by formulæ (73) and (74.) We have

$$\text{Sin. C} = \frac{\frac{1}{2}\ ch}{r} \text{ and tan. } \tfrac{1}{2}\ \text{C} \cdot \tfrac{1}{2}\ ch = \text{D}\ d = 10 \cdot 10$$

| | | | |
|---|---|---|---|
| $r$ | = 300 feet | co. ar. | log. = 7·5228787 |
| $\frac{1}{2}\ ch$ | = 10 " | | log. = 1·0000000 |
| C | = 1° 54′ 36″·767 | | sin. = 8·5228787 |

| | | | |
|---|---|---|---|
| ½ C | $= 0° 57' 18''·383$ | | tan. = 8·2219693 |
| ½ ch | = 10 feet | | 1·0000000 |
| D d | = (10) . (10) = 0·16671 feet | | 9·2219693 |
| $10^2\ y$ = | 0·16666 | | |
| Difference | = 0·00005 | | |

It will be seen, by comparing D *d*, with $10^2\ y$, that the errors of the approximate formula are too small to be noted in the practical operations of curving rails.

It will be noticed that the above ordinates extend from the tangent line D 10, D 10″, to the curve A B D. It frequently happens that it will be more convenient to set out the curve from the chord line A *d* B, which line may be readily represented upon the pattern board by straining a small string or wire from A to B. To prepare ordinates to be thus used, we subtract the ordinates found successively from the greatest ordinate, (which, in our example, is = $10^2\ y$.) Thus,

| | | | |
|---|---|---|---|
| $10^2\ y - 0$ | = D *d* | = | 0·1666666 |
| $10^2\ y - y$ | = *a* 1′ | = | 0·1650000 |
| $10^2\ y - 2^2\ y$ | = *b* 2′ | = | 0·1500000 |
| $10^2\ y - 3^2\ y$ | = *c* 3′ | = | 0·1516666 |
| $10^2\ y - 4^2\ y$ | = *d* 4′ | = | 0·1400000 |
| $10^2\ y - 5^2\ y$ | = *e* 5′ | = | 0·1250000 |
| $10^2\ y - 6^2\ y$ | = *f* 6′ | = | 0·1066666 |
| $10^2\ y - 7^2\ y$ | = *g* 7′ | = | 0·0851111 |
| $10^2\ y - 8^2\ y$ | = *h* 8′ | = | 0·0600000 |
| $10^2\ y - 9^2\ y$ | = *i* 9′ | = | 0·0316666 |
| $10^2\ y - 10^2\ y$ | = *j* 10′ | = | 0·0000000 |

Before we leave this subject, we would remark, that all curves of a less radius than 3000 feet, provided they are laid with rails twenty-one feet in length, should be curved; and a board prepared as a pattern will facilitate, and add both to the convenience and accuracy of the operations.

**(57)** It frequently becomes necessary for the engineer to ascertain the radius of a small part of a curve in a railroad track; as, for example, when called upon to lay down a side track, a turnout, or to connect a branch road with a curve, the radius of which is unknown. Many formula may be deduced for the solution of this problem, each possessing nearly equal convenience; we shall, however, limit our investigations to some two or three of those in common use.

Let T′ A T represent a portion of a curve, the radius of which it is desirable to ascertain. We measure the chords A T and A T′, each of the same length, which we represent by *c;* then ascertain the point B in the middle of the chord T′ T; and measure the ordinate A B, which we represent by *b*. Then, representing the radius of the curve by *r;* and the radius of the tables by R; we have, in the triangle A B T′, $c : R :: b : \cos. A = \frac{b}{c}$; and

$$\text{Cos. } A : \tfrac{1}{2}\, c :: R : r = \frac{\frac{1}{2}\, c}{\cos. A} = \frac{\frac{1}{2}\, c^2}{b} \qquad (75)$$

For an example of computation, we will suppose $c = 100$ feet; and $b = 8$ feet. Then,

| | | | |
|---|---|---|---|
| $c^2 = 100^2$ | | log. | $= 4{\cdot}0000000$ |
| $b = 8$ | co. ar. | log. | $= 9.0969100$ |
| 2 | co. ar. | log. | $= 9{\cdot}6989700$ |
| $r = 625$ feet | | log. | $= 2{\cdot}7958800$ |

Again, let T′ A T represent, as before, a portion of a curve, the radius of which we desire to ascertain; let the chord T′ T be represented by $a$; and let $a = 199 \cdot 359$ feet; and $b = 8$ feet; as above. We then, by dividing $a$ by 2, have in the triangle A B T′, (taking $b$ as a cosine, and $\frac{1}{2}$ $a$ as a sine,) the following analogy,

$$\text{Cos.} : \text{sin.} :: \text{R} : \text{tan. A} = \frac{\frac{1}{2} a}{b} \tag{76}$$

$$\text{Then, Sin. A} : \tfrac{1}{2} a :: \text{R} : c = \frac{\frac{1}{2} a}{\text{sin. A}}$$

$$\text{Cos. A} : \tfrac{1}{2} c :: \text{R} : r = \frac{\frac{1}{2} c}{\text{cos. A}} = \frac{\frac{1}{2} a}{2 \text{ sin. A cos. A}} \tag{77}$$

EXAMPLE OF CALCULATION.

| | | | |
|---|---|---|---|
| $\frac{1}{2} a$ | = 99·6795 | | log. = 1·9986059 |
| $b$ | = 8 | co. ar. | log. = 9·0969100 |
| A | = 85° 24′ 45″·17 | | tan. = 1·0955159 |
| A | = 85° 24′ 45″·17 | co. ar. | sin. = 0·0013942 |
| A | = | co. ar. | cos. = 1·0969102 |
| | 2 | co. ar. | log. = 9·6989700 |
| $\frac{1}{2} a$ | = 99·6795 | | log. = 1·9986059 |
| $r$ | = 625 feet | radius | log. = 2·7958803 |

Again, let T′ A T represent the segment of the curve whose radius is desired, and the half chord T′ B $= a$; the ordinate A B $= b$; and the chord T′ A $= c$; and the diameter A D $= d$; the radius $= r$. We here remark, that the angle A is common to the triangle T′ A B, and the triangle T′ A D; and the angle B in the triangle T′ B A, and the angle T′ in the triangle D T′ A are each a right angle; consequently, the two triangles are similar. We now have

$$a^2 + b^2 = c^2$$

$$\text{and } b : c :: c : d = \frac{c^2}{b} = \frac{a^2 + b^2}{b} = \frac{a^2}{b} + b = 2\,r \tag{78}$$

Example of Computation. Let $a = 99 \cdot 6795$ feet; $b = 8$ feet. Then,

[FIG. 25.]

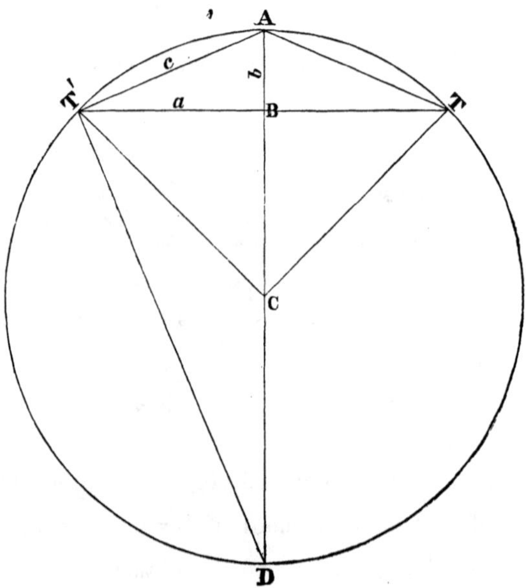

| | | | |
|---|---|---|---|
| $a^2$ | $= \overline{99 \cdot 6795}^2$ feet | | log. $= 3 \cdot 9972118$ |
| $b$ | $= 8$ feet | co. ar. | log. $= 9 \cdot 0969100$ |
| | 1242 | | log. $= 3 \cdot 0941218$ |
| $+ b$ | $= 8$ | | |
| | $2 \overline{)1250}$ | | |
| $r$ | $= 625$ | | |

We have here thus endeavored to apply the same elements in each of our examples of computations, by way of testing the different methods, and we find each of them to give the same results.

(**58**) Before we take our final leave of railroad tracks, we will add a formula for elevating the outside rail of curves. We extract what we shall say upon the subject from Article XV. of the 22d vol. of the *American Journal of Science*, 1832.

The article was written by J. Thompson, (Engineer, and late Professor of Mathematics in the University of Nashville, Tenn.,) and commences with a discussion of a formula given in Colonel Long's work on railroads, which he shows to be erroneous. With this criticism it is not our intention to meddle; but, as Mr. Thompson has, in the course of his remarks, developed a convenient and an accurate formula, we shall endeavor to extract only so much as may seem to be connected with its explanation.

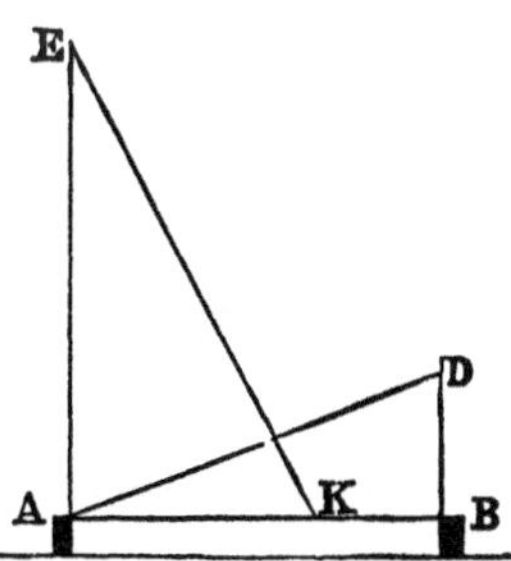

"Let C A B represent a horizontal surface on which a railway is situated; A and B the rails placed in a circular curve around C as a centre. A car moving over the rails A and B, around the centre C, will be acted upon by two forces, one horizontal and centrifugal, arising from the motion of the car in a curved line, and acting in a direction from the centre C; the other, the force of gravity, acting in a vertical direction. I omit here, as not necessary in the present investigation, the moving force derived from animal or other power acting in a direction of a tangent to the curve. Let the horizontal line A K represent the centrifugal force above mentioned, and the line E A the force of gravity. It is evident that the resultant of these two forces will be E K, which will represent both the *intensity* and the *direction* of the pressure of the loaded car upon the rails. The line E K, therefore, representing the direction of pressure, the rails should be so placed that this line may be perpendicular to the plane passing through them. Draw the vertical line B D, and through A draw A D, perpendicular to E K; B D will be the elevation of the exterior rail above the interior, and the angle D A B will be the inclination of the plane of the rails to the horizon. The centrifugal force A K, compared with the force of gravity A E, is easily found, when the radius of curvature of the track and the velocity of the car are given. The distance between the centre C, and the middle of the track, may be considered as the radius of curvature.

"We may obtain a very simple algebraic expression for the elevation of the exterior rail. Let $g =$ force of gravity; $c =$ centrifugal force; $d =$ distance between the rails; and $E =$ required elevation; R and V representing radius and velocity. Then, by the similar triangles E A K and A B D, we have $E = \frac{c\,d}{g}$; but, by central forces, $c = \frac{V^2}{R}$; hence, $E = \frac{d\,V^2}{R\,g}$ in this expression; $g$ is always a constant quantity, and equal to 32·2 feet.

"If the velocity of a car on a railway were always the same, we should have no difficulty in assigning the proper elevation of the exterior rail. But, as there must be necessarily a great variety in rates of travelling, an elevation for a rate of twenty miles per hour would be much too great for a rate of eight, twelve, or fifteen miles per hour. Perhaps the elevation required by the *mean velocity* would be most eligible.* There is one view of the subject, however, which ought to be taken into consideration in the location of the exterior rail. When a car moves with great velocity on a curved road, and the planes of the rails are horizontal, the flange of the fore wheel on the exterior rail is exposed to very great friction, which operates as a retarding force, and injures both the car and the railway; this friction is diminished, though not altogether removed, by giving the exterior rail the elevation which the velocity and radius require. In order to reduce the friction still further, or remove it altogether, it would perhaps be advisable to increase by a small quantity the elevation obtained as above.† It is evident that a car moving on the inclined plane A D, will tend by its own weight to approach A, and recede from D; this will oppose the centrifugal force by which the flange is pressed against the rail D, and thus the friction will be in whole or in part removed. I know it has been maintained that the flange of the hind wheel on the interior rail produces as much friction as the flange of the exterior

* It has been the practice of the author of the foregoing papers to elevate the exterior rail to suit the highest velocity with which the regular trains are supposed to run over the curve. At the present time we should not think thirty-five miles the hour too great. Quick trains produce greater friction upon the exterior rail, and are more liable to accident than slow trains.

† It has been the practice of the author to add one fourth of an inch to the computed elevation.

fore wheel. It may, however, be shown, from various considerations, that if either of the hind wheels produces friction, it is rather the exterior one; indeed, we may suppose that motion is communicated to the hind wheels by a force which acts precisely in the same direction as if they were moved by animal power, the direction being nearly a tangent to the curve. This being admitted, the flanges of the two exterior wheels sustain all the friction occasioned by curvature. It may be observed, however, that when the distance between the fore and the hind wheels is comparatively very great, the direction of the force moving the hind wheels will vary considerably from the tangent, and consequently the friction will be diminished."*

* Although we agree with Mr. Thompson in the main, we do not fully agree with his concluding remarks. Mr. Thompson says, "It may, however, be shown, from various considerations, that if either of the hind wheels produces friction, it is rather the exterior one; indeed, we may suppose that motion is communicated to the hind wheels by a force which acts precisely in the same direction as if they were moved by animal power, the direction being nearly a tangent to the curve. This being admitted, the flanges of the two exterior wheels sustain all the friction occasioned by the curvature. It may be further observed, however, that when the distance between the fore and hind wheels is comparatively very great, the direction of the force moving the hind wheels will vary considerably from the tangent, and consequently the friction will be diminished."

The reasoning of Mr. Thompson, doubtless was applicable to cars sustained upon two axles and four wheels only, one axle being situated near the forward end of the car, and the other near the back or hind end. Now, if the car be short, the axles must of course be near each other; in this condition, the flanges of both the forward and hind wheels may grind the exterior rail, the forward wheels of course grinding much the hardest. A distance, however, between the axles can be readily ascertained which will relieve the hind wheels from the friction of the flanges against either the interior or exterior rail; then, expanding the distance between the axles, the flanges of the hind wheel will begin to grind against the interior rail, and the greater the distances between the axles the greater will be the friction. I would observe, however, that the forward wheel flange will, under all distances between the axles, grind upon the exterior rail, and will grind more and more severely in proportion as the distance between them increases. These notions are based upon the condition that the axles are firmly and permanently secured to the car, and at right angles with its frame. We might easily demonstrate the position we have here taken by diagrams if it were thought necessary, but the change produced by the adaptation of what we term four-wheel trucks to our long car bodies, which permits

We have thus copied Mr. Thompson's article, with his remarks, omitting only the portion relating to Mr. Long's formula.

EXAMPLE OF COMPUTATION. Assuming a radius, R = 4000 feet; and a velocity of 35 miles per hour; 35 miles per hour = V = 51·33 feet per second of time; the width between the rails being = $d$ = 4·7 feet.

$$\text{Formula } E = \frac{d\,V^2}{R\,g}$$

| | | | |
|---|---|---|---|
| R | = 4000 feet | co. ar. | log. = 6·3979400 |
| $g$ | = 32·2 | co. ar. | log. = 8·4921441 |
| $V^2$ | = $\overline{51 \cdot 33}^2$ | | log. = 3·4207990 |
| $d$ | = 4·7 | | log. = 0·6720979 |
| E | = 0·09615 | | log. = 8·9829810 |

Again, for the purpose of showing the changes of E, consequent upon the changes of R, we assume R = 2000 feet; the other expressions remaining the same. Thus,

| | | | |
|---|---|---|---|
| R | = 2000 feet | co. ar. | log. = 6·6989700 |
| $g$ | = 32·2 | co. ar. | log. = 8·4921441 |
| $V^2$ | = $\overline{51 \cdot 33}^2$ | | log. = 3·4207990 |
| $d$ | = 4·7 | | log. = 0·6720979 |
| E | = 0·1923 feet | | log. = 9·2840110 |

We believe we have in the foregoing pages examined every distinct species of curve that enters into the construction of a railroad. We were aware, as we proceeded in our investigations,

the arrangement of the axles to a very near approximation with the radii of the curve, by the force of the wheel flanges against the exterior rail, seems to render such an undertaking unnecessary, particularly as the railroad companies appear to be universally adopting them. But, it is not our intention to discuss generally the principles which should govern the construction of cars. Having concluded to adopt the formula of Mr. Thompson, we thought, in justice to him, we were bound to copy his remarks. We would only further mention, respecting Mr. Thompson's remarks, that we cannot discover any material difference in the action of the flanges of the wheels upon the curved rails, whether the car receives its motion from a force pulling in front or pushing behind.

that many modifications of the formula we have deduced would frequently be called for; but, as we have before stated, it has not been our intention to exhaust the subject, but merely to give a formula for the most prominent of each class, or rather for that class which most frequently present themselves to the engineer while engaged in construction.

We contemplated, when we commenced our work, closing our paper here; but it has occurred to us, that the inexperienced engineer might feel the want of some convenient plan or system of computing the cubic contents of excavations and embankments. For the purpose of supplying those wants, we add the following.

An investigation of formulæ for the computation of the cubic contents of earth, excavations, embankments, masonry, etc., in constructing railroads.

(**59**) An article quoted from *Silliman's Journal*, by Professor Eaton, states, in effect, as follows; that whereas the sections into which the engineer would divide the excavations upon a railroad, readily admit of being subdivided into pyramids, wedges, and parallelopipeds; therefore, if you add the area of both ends of the section, to four times its middle area, divide the sum by six, and multiply the quotient by the length of the section, the product will give its solid contents.

This problem can be readily demonstrated to be strictly correct, provided the sides of the section are perfect planes. It is the constant endeavor of every skilful engineer so to arrange the sections that, were the irregularities of the earth to be pared down, so as to produce regular planes between the points which he has chosen to take his levels at, the solid contents of the earth contained in the section would be just sufficient to fill the hollows, and make the surfaces planes between them.

**(60)** In Professor Eaton's enunciation of the rule above quoted, I did not discover any method of ascertaining the middle area of the section; it being evident that an arithmetical mean of the areas of the ends of the section would not uniformly produce the desired result. To supply this deficiency in the formula is the main object of the present paper; but, as the original formula may not be within the reach of every individual who may feel interested in seeing an investigation and demonstration of it, we have thought a brief investigation might not be out of place; besides, it will aid much in rendering the subject more plain and intelligible. I shall, however, take it for granted that the interested reader will know enough of geometry to be familiar with the common formulæ for measuring the solid contents of pyramids, parallelopipeds, wedges, and surfaces of the cross sections of the excavations of a railroad; we shall, therefore, only allude to the most common formulæ for measuring solids and superficies, as we may have occasion to compare them with the formula to be deduced.

Commencing with the pyramid.

**(61)** According to the rule, we have to add the area of the base of the pyramid to four times the area of its middle section, (taken parallel with said base,) divide the sum by six, and multiply the quotient by the height of the pyramid; the product will give its solid contents.

[FIG. 26 (1).]

THE PYRAMID.

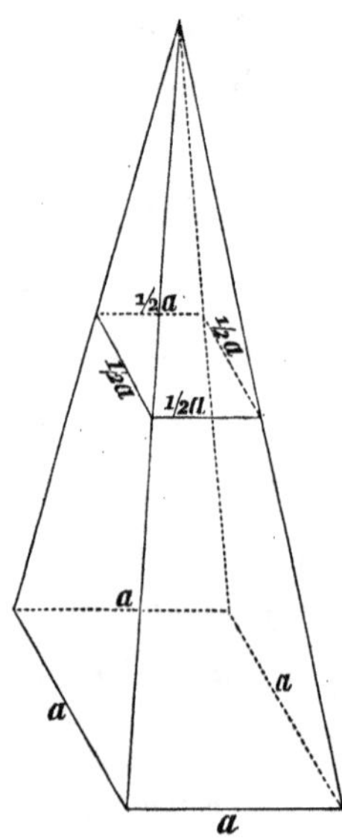


[FIG. 26 (2).]

THE WEDGE.

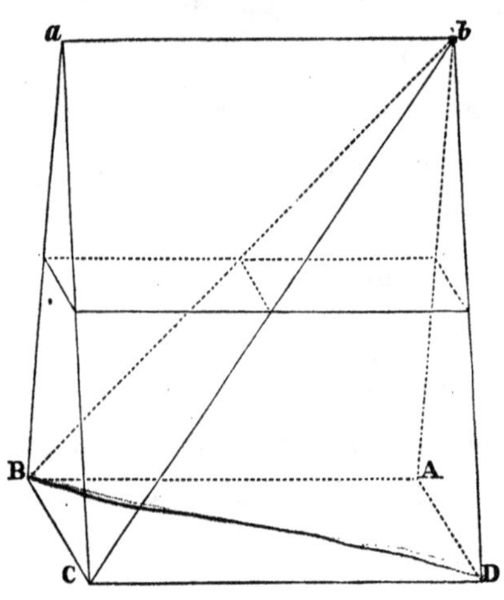

For the purpose of proof, or demonstration, let us suppose a pyramid of four sides, Fig. 1, the angles of its base being right angles, and the lines circumscribing said base of equal length, which length we represent by $a$; it is evident that the lines which circumscribe the middle section of said pyramid (taken parallel with said base) will be equal to $\frac{1}{2}a$; if then we represent the height by $h$, we have for the area of the base $a \times a = a^2$; for the area of the middle section $\frac{1}{2}a \times \frac{1}{2}a = \frac{1}{4}a^2$; then, by the rule, $\left(\frac{a^2 + \frac{1}{4}a^2\,4}{6}\right)h = \frac{2a^2h}{6} = \frac{1}{3}a^2h =$ the solid contents; which corresponds exactly with the formula in common use (1.)

This equation shows the area of the middle section of a pyramid to be one fourth of the area of the base; and this proposition is universal and equally correct in every species of pyramid, whether it be three, four, or many-sided, regular or irregular.

Secondly, the wedge.

**(62)** We next apply the rule to the measurement of the wedge. Let us now suppose a regular or symmetrical wedge, with a base circumscribed by lines of equal length, which we represent by $a$; it is obvious that of the lines which circumscribe the middle section of the wedge, (taken parallel with the base,) two of them will equal $a$; and the other two will equal $\frac{1}{2}a$. Then, making, as before, $h$ equal the height, or length; we have, by the rule,

For the area of the base $a \times a = a^2$;
For middle section $a \times \frac{1}{2}a = \frac{1}{2}a^2$.

Then, $\left(\frac{a^2 + \frac{1}{2}a^2\,4}{6}\right)h = \frac{3a^2h}{6} = \frac{1}{2}a^2h =$ the solid contents; which corresponds with the formula in common use for determining the solid contents of the wedge (2.)

From this equation we learn, that the area of the middle section is equal to one half the area of the base. By cutting the wedge into pyramids we may compute its solid contents in a way somewhat different, with the same formula.

Let A B C D, in Figure 2, represent the base end of the wedge; and *a b* the edge or sharp end; cut the wedge diagonally through the plane *b* B C. The wedge is thus divided into two pyramids; A B C D *b*, and *a b* B C; the pyramid A B C D *b* being a four-sided one; and the pyramid *a b* B C being a triangular or three-sided one. We may further cut the four-sided pyramid in the plane *b* B D, which divides that pyramid into two three-sided pyramids. The wedge will then consist of three triangular pyramids; but, as the same rule for determining the cubes applies to three, four, and many-sided pyramids, we shall, in our further investigations of the mensuration of the wedge, only use the four-sided pyramid, in connection with the blind pyramid *a b* B C; (this pyramid is so named because it presents no area in either of the surfaces of the cross section of a cut in a railroad excavation.) Pyramids of this character enter into the calculations of nearly every cross section, and my principal object in introducing it in the wedge, is, for the purpose of testing the method of computing its solid contents.

If we now compare the solid contents of the four-sided pyramid, it will be observed that we make use of the whole area of the base of the wedge, and add thereto four times the quarter area of the base; which quarter is equal to one half the area of the middle section of the wedge; this sum, divided by six, and the quotient multiplied by the length of the wedge, gives the solid contents of the pyramid. Then, to complete the measurement of the wedge, according to the rule; having divided the area of the base, plus four times one fourth of the area of the base, etc.; as explained before, the remainder of the area of the middle section will of course be equal to one fourth of the area of the base; which, multiplied by four, its product divided by six, and the quotient multiplied by the length of the wedge, the operation will be complete; (and have been performed in a different method,) the result being the same as in equation (2.)

I would however mention, that we found by equation (2) that the middle area of the wedge was equal to half the area of the base; and, in measuring the four-sided pyramid, the middle area of which is equal to one quarter of the area of the base, of course it is equal to one half of the middle area of the wedge, leaving the other half for the middle area of the blind pyramid.

To elucidate this, let us introduce the calculations.

Using the former notations, we have, in the mensuration of the four-sided pyramid, for the base $a \times a = a^2$, the middle area $\frac{1}{2} a \times \frac{1}{2} a = \frac{1}{4} a^2$; then, $\left(\frac{a^2 + \frac{1}{4} a^2 4}{6}\right) . h = \frac{2 a^2 h}{6}$ (3.)

In the mensuration of the blind pyramid *a b* B C, we have for the middle area, (*a b* being equal to *a*, and B C being also equal to *a*) $\frac{1}{2} a \times \frac{1}{2} a = \frac{1}{4} a^2$; and $\left(\frac{\frac{1}{4} a^2 4}{6}\right) . h = \frac{a^2 h}{6}$ (4.)

If we now add the results of the above equations, (3 and 4,) their sum will be equal to the contents of the wedge, as found in equation (2;) thus proving that the mensuration of the blind pyramid is exact, according to the rule. As a further proof, we may

[Fig. 27.]

In which the two separate Diagrams referred to by the text may be traced

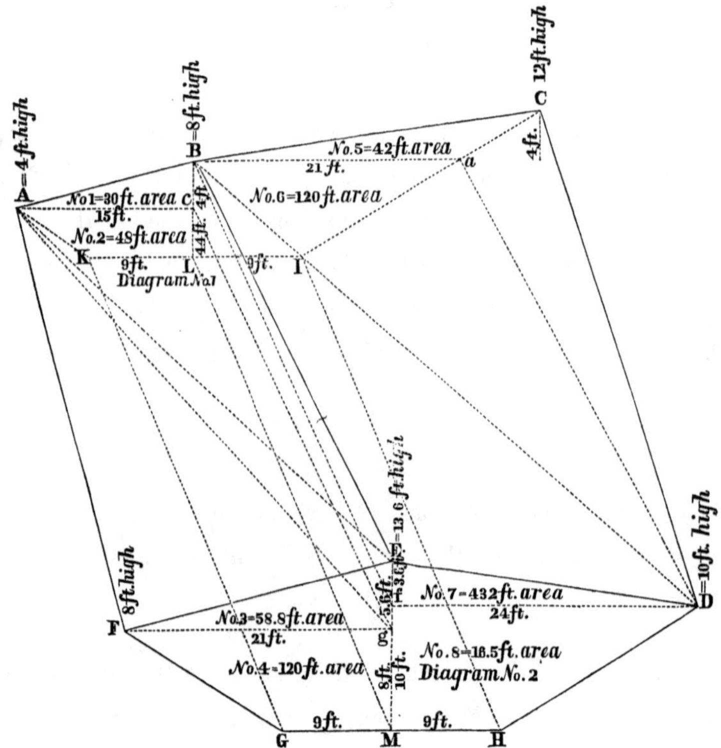

now measure the blind pyramid in this manner. Making $a$ B C the base, and $a\ b$ the height, which we shall denote by $h$; then, according to rule, and using the same notation, we have

$$\left(\frac{\frac{1}{2}\,a\,h + (\frac{1}{4}\,a \times \frac{1}{2}\,h)\,.\,4}{6}\right).\ a = \frac{a^2\ h}{6}\ (5;)$$

the same as in equation (4.)

We might produce a great variety of proofs to show the accuracy of the rule. Thus, if we should endeavor to measure a four-sided square parallelopiped by the rule, the simplest manner of proceeding will be to divide it into two wedges, and then apply the rule; or, we may cut off four pyramids, leaving a large blind pyramid, which, in order to determine its contents, will require that we should determine the length of the diagonals, (this method will be exact only when the diagonals are at right angles), and then multiplied into each other, will give four times the middle area required; this, divided by six, and multiplied by the length, will give the solid contents of the blind pyramid.

But, this last method is somewhat complicated, inasmuch as we should be obliged to find the diagonals by extracting their square roots from the sum of the squares of the other sides; or, we may find the diagonal by trigonometry. I shall, therefore, only give an example of determining the cubic contents of the parallelopiped, by dividing, first, into two wedges. This example we give merely as an illustration of the method.

Retaining the former notation, we have for the measurement of the apparent pyramids, the bases and four times the middle area of each pyramid, equal to twice the area of the bases. See equation (1,) which, for both pyramids, is equal to $4\ a^2$, and four times the middle area of one of the blind pyramids, will be $a \times a = a^2$ each;

there being two of them, we have, therefore, $4\ a^2 + 2\ a^2 = 6\ a^2$, and $\frac{6\ a^2\ h}{6} = a^2\ h$; the same result as by the ordinary method of determining the cubic contents of a parallelopiped.

Of course the above is not the most convenient method, but will serve to show the application of the rule to the measurement of the frustrum of a pyramid.

The rule is also peculiarly applicable to the measurement of wedges in which one end is wider than the other, and to almost every figure imaginable which is bounded by right lines and plane surfaces.

We will now give the calculations of a few imaginary figures of different forms, after the manner of our practice, to determine the cubic contents of excavations, embankments, masonry, etc.

**(63)** Let A B C I L K represent one end of the supposed section, which we will denominate No. 1, (in Fig. 27,) and D E F G M H, the other end, which we denominate No. 2. After preparing diagrams of the ends of the sections, and marking the heights of the points A B C of diagram No. 1, and of D E F in No. 2, we then divide the diagrams into figures which we shall now proceed to describe.

Firstly, we divide No. 1 by the perpendicular line L B, the point L representing the centre of the road bed.

Secondly, draw the line A *c* parallel to the base, or the road bed, K I ; which is always level.

Thirdly, draw the line B *a* parallel to the line K I; then the diagram will be divided into the triangles A B *c* and B C *a*, and the trapezoids A K L *c* and B L I *a*. We now propose for the dimensions that K L and L I shall each be 9 feet long, and that the height of A be 4 feet, the height of B 8 feet, the height of C 12 feet. In order to determine the areas of the above mentioned trapezoids and triangles, we first determine the length of the lines A *c* and B *a*; the length of A *c* being equal to the height of A, plus one half the height of A + K L, upon the supposition that the slopes of the cuttings are as three to two. Thus,

| | |
|---|---|
| The height of A . . . . . . . . . . | = 4 feet |
| ½ " " . . . . . . . . . . | = 2 |
| K L . . . . . . . . . . . | = 9 |
| Length of A *c* . . . . . . . . . . | = 15 feet |

which we mark upon diagram No. 1.

To find the length of B *a*, we have the height of B L + ½ the height of B L + L I = B *a*. Thus,

| | |
|---|---|
| The height of B L . . . . . . . | = 8 feet |
| ½ " " . . . . . . . . . . | = 4 |
| Length of L I . . . . . . . . | = 9 |
| Length of B *a* . . . . . . . . | = 21 feet |

which we mark upon the diagram also.

Having prepared the diagram, we proceed to determine the area of the trapezoid A K L *c*. We have found

| | |
|---|---|
| A *c* . . . . . . . . . . . . | = 15 feet |
| K L . . . . . . . . . . . . | = 9 |
| A *c* + K L . . . . . . . . . . | 2)24 |
| Mean, or ½ (A *c* + K L) . . . . . . | = 12 |
| Height of A above K = L *c* . . . . . . | = 4 |
| Area . . . . . . . . . . . | = 48 feet |

To find the area of the trapezoid *a* B L I, we have found

| | | |
|---|---|---|
| *a* B | = | 21 feet |
| L I | = | 9 |
| *a* B + L I) | = | 2)30 |
| ½ (*a* B + L I) | = | 15 |
| Height of B | = | 8 |
| Area | = | 120 feet |

To find the area of the triangle A B *c*, we have found

| | | |
|---|---|---|
| A *c* | = | 15 feet |
| ½ difference of heights B and A | = | 2 |
| Area | = | 30 feet |

To find the area of the triangle *a* B C, we have found

| | | |
|---|---|---|
| B *a* | = | 21 feet |
| ½ difference of heights of C and B | = | 2 |
| Area | = | 42 feet |

Having determined the areas of each figure composing diagram No. 1, and marked the same upon it, we then proceed to divide diagram No. 2 into triangles and trapezoids, and compute their areas in a manner similar in principle to that adopted in No. 1.

Assuming the height of F = 8 feet; the height of E = 13·6 feet; the height of D = 10 feet; and G M and M H = 9 feet in length each. Firstly, we have, in the triangle F *g* E,

| | | |
|---|---|---|
| The length of the line F *g*, equal the height of F | = | 8 feet |
| + half the height of F | = | 4 |
| + G M | = | 9 |
| Wherefore, F *g* | = | 21 feet |

We then have half the difference in heights of *g* and E = half the difference of F and E. Thus,

| | | |
|---|---|---|
| The height of F | . . . . . . . . . . | = 8 feet |
| " " E | . . . . . . . . . . | = 13.6 |
| Difference | . . . . . . . . . . | 2)5.6 |
| Half difference of g E | . . . . . . . . | = 2·8 feet |

We now find the area of the triangle F $g$ E = 21 × 2·8 = 58·8 feet.

Secondly. In the triangle D E $f$ we have

| | | |
|---|---|---|
| The length of the line D $f$ equal the height of D | . | = 10 feet |
| + " " ½ D | . | = 5 |
| + M H | . . . . | = 9 |
| Wherefore $f$ D | . . . . . . . . . | = 24 feet |

Then we have half the difference in height of $f$ and E equal to half the difference of D and E; thus,

| | | |
|---|---|---|
| The height of D | . . . . . . . . . . | = 10·00 feet |
| " " E | . . . . . . . . . . | = 13·60 |
| Difference | . . . . . . . . . . | 2) 3·60 |
| Half difference | . . . . . . . . . . | = 1·80 |

We now find the area of the triangle D E $f$ = 24 × 1·8 = 43·2 feet.

Thirdly. To find the area of the trapezoid F G M $g$; we found in the foregoing,

| | | |
|---|---|---|
| F $g$ | . . . . . . . . . . | = 21·00 feet |
| G M | . . . . . . . . . . | = 9·00 |
| F $g$ + G M | . . . . . . . . | 2)30·00 |
| Mean length | . . . . . . . . . | = 15·00 |
| Then, M G = height of F | . . . . . . | = 8·00 |
| Area | . . . . . . . . . . | = 120·00 feet |

FIG. 28.]

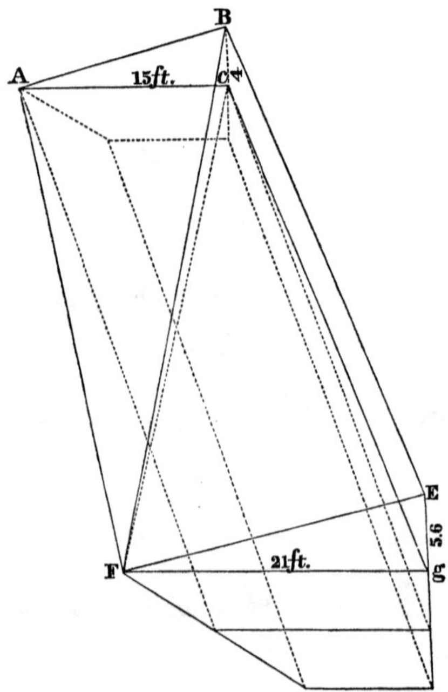

Fourthly. Then, to find the area of the trapezoid D H M *f*; we find in the foregoing,

| | |
|---|---|
| D *f* . . . . . . . . . . . | = 24·00 feet |
| M H . . . . . . . . . . . | = 9·00 |
| D *f* + M H . . . . . . . . | 2)33·00 |
| Mean length . . . . . . . . | = 16·5 |
| Then, M *f* = height of D . . . . . . | = 10·00 |
| Area . . . . . . . . . . . | = 165·00 feet |

Having ascertained the several areas of the divisions of the cross sections, and marked the same upon the diagram, our next operation will be to ascertain the cubic contents of that portion of the half section lying above the lines F *g*, *g c*, *c* A, and A F.

**(64)** It will be evident, from an inspection of the drawing, that the upper portion of the half section may be divided into two apparent, and one blind pyramid. This solid admits of two distinct methods, or plans of divisions, so that, if the surface is not twisting, it matters not which of the plans is adopted. For the purpose of a test, we will consider the divisions under both aspects; firstly, as represented in the drawing of the section, (viz., Fig. 27;) and secondly, as in Fig. 28. Whatever difference there may be, if any, will be seen in the different dimensions the blind pyramid will assume; hence we will confine ourselves to the comparison of the contents of this solid.

Firstly, we have A *c* × *g* E = 4 times the centre area of the blind pyramid; and, in the second form, we have F *g* × *c* B = 4 times the centre area, as above.

| First Form. | | Second Form. | |
|---|---|---|---|
| A *c* . . . . . . | = 15· feet | F *g* . . . . . | = 21·00 feet |
| *g* E . . . . . | = 5·6 | *c* B . . . . . | = 4·00 |
| | 9·0 | 4 times middle area | = 84·00 |
| | 75· | | |
| 4 times middle area . | = 84·00 | | |

[Fig. 29.]

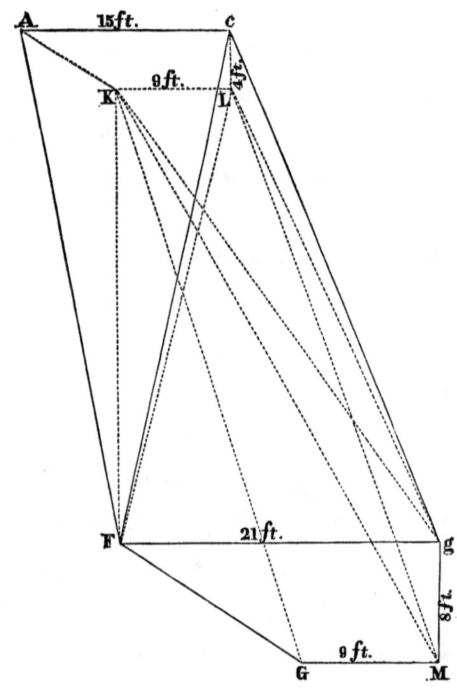

It thus appears that the top surface is a perfect plane: had it been warped, or twisted, the middle areas of the blind pyramid under both aspects would not have been alike.

We have now, in the remainder of the first half section, the solid A K L $c$ $g$ F G M. The lines F $g$, G M, K L, A $c$, being parallel, the solid admits of a division into two wedges; and these wedges, as we have already shown, may be divided into one apparent and one blind pyramid each; the area F $g$ M G being the base of one of the apparent pyramids, and the area A $c$ L K being the base of the other. The surfaces of this solid being perfect planes, it matters not in what manner we form the wedges, and cut from them the blind pyramids, as their combined measurements will be the same. For example, if the wedges be formed by cutting the solid through the plane A $c$ M G, four times the middle area of one of the blind pyramids will be equal to M $g$ × A $c$, and four times the middle area of the other will be equal to L $c$ × G M. Or, we may form the wedges by cutting the solid through the plane K L $g$ F; then, four times the middle area of one of the blind pyramids will be equal to L $c$ × F $g$, and the other will be equal to $g$ M × K L. Before we proceed further with our investigation, we will compare the contents of the blind pyramids as ascertained by both methods of division.

Firstly. We have A $c$ × M $g$ = 4 times the middle area of one of the blind pyramids.

| | |
|---|---|
| A $c$ . . . . . . . . . . . . . | = 15 feet |
| M $g$ . . . . . . . . . . . . . | = 8 |
| 4 times the middle area, . . . . . . . . | = 120 feet |

Again, we have L $c$ × G M = 4 times the middle area of the other blind pyramid. Thus,

| | | |
|---|---|---|
| L $c$ . . . . . . . . . . . . . | = | 4 feet |
| G M . . . . . . . . . . . . . | = | 9 |
| 4 times the middle area . . . . . . . . | = | 36 |
| Then add the area found above . . . . . | = | 120 |
| Gives the sum of the middle areas of both . . | = | 156 |

Secondly. We have L $c$ × F $g$ = 4 times the middle area of one of the blind pyramids. Thus,

| | | |
|---|---|---|
| L $c$ . . . . . . . . . . . . . | = | 4 feet |
| F $g$ . . . . . . . . . . . . . | = | 21 |
| 4 times the area . . . . . . . . . | = | 84 feet |

Again, we have M $g$ × K L = 4 times the middle area of the other blind pyramid. Thus,

| | | |
|---|---|---|
| M $g$ . . . . . . . . . . . . . | = | 8 feet |
| K L . . . . . . . . . . . . . | = | 9 |
| 4 times the area . . . . . . . . . | = | 72 |
| Add the area found above . . . . . . | = | 84 |
| Sum of middle areas of both . . . . . . | = | 156 feet |

Hence we see that the computations prove that the final result will be the same under both methods of computation.

We will now sum up the measurement of the supposed divisions of the first half section.

| | | | |
|---|---|---|---|
| 1st. | We have the area of the base of the triangle F $g$ E A = 58·8 feet, and + 4 times its middle area = 58·8;* (See Fig. 27) . . . . . . . . . | = | 117·6 feet |
| 2d. | We have the area of the base of the pyramid A B $c$ E = 30 feet, and + 4 times its middle area = 30 . . . . . . . . . . . . . . | = | 60·0 |
| 3d. | We have 4 times middle area of blind pyramid A $c$ E $g$ . . . . . | = | 84·0 |
| 4th. | We have the area of the base of the pyramid F G M $g$ A = 120 feet, and + 4 times the middle area = 120 . . . . . . . . . . . . | = | 240·0 |
| 5th. | We have the area of the base of the pyramid A $c$ L K $g$ = 48 feet, and + 4 times the middle area = 48 . . . . . . . . . . . . . | = | 96·0 |
| 6th. | We have the sum of the middle areas of the remaining two blind pyramids | = | 156·0 |
| | | | 753·6 |
| 7th. | Taking the length of the section = 100 feet . . . . . . . | | 100 |
| | Dividing by . . . . . . . . . . . . . . . . | | 6)75360·0 |
| | Solid contents . . . . . . . . . . . . . . | = | 1256.0 |

**(65)** It may not be amiss here to remark, that we always, when the nature of the case will admit, divide each section into two parts by a plane passing through the vertical lines M E and B L, and compute each portion separately, (see Fig. 27.) The reason for this is, that a level is always taken over the centre of the road bed in every cross section, and, in a majority of the cases which occur, there are only two other levels taken; viz., one at the right, and one at the left hand slope stakes; and whenever it becomes necessary to take other levels at the right or left of the centre, it will still be convenient to preserve the centre division. And it will be apparent when we have completed our computations, that our system of dividing the cross sections into triangles and trapezoids, is pecu-

* In the early part of this discussion, we proved that the middle area of a pyramid, taken parallel to its base, was equal to one fourth of the area of its base; hence, 4 times the middle area will be equal to that of the base.

liarly adapted to this method of computing cubic contents, and affords a very convenient, as well as an accurate method of ascertaining the area of said cross section.

Before we proceed to the computation of the cubic contents of the remaining, or second half section, we would remark that the portion of the second half section which lies above the plane passing through B *f* D *a*, having its upper surface much twisted or warped, admits of two forms of division, which will be found by computation to give different results; one of which will be applicable to one form of surface, and the other to another. And that the portion of the half section lying below the plane B *f* D *a* having all of its surfaces perfect planes, the computation will give correct, and of course like results, from whichever of the forms the divisions may take.

We now proceed to the examination of the upper portion of said half section.

Firstly. If we suppose the upper surface to have this form, viz., that of a plane through D E B, and intersecting in the line D B, a plane passing through D B C, the solid will then contain only two apparent pyramids, with no blind pyramid; viz., the apparent pyramids D *f* E B and B *a* C D; the measurement of which is as follows:

| | | |
|---|---|---|
| For the area of base of D E *f* B = 43·2 feet, and + 4 times middle area = 43·2 | = | 86·4 feet |
| " " " *a* B C D = 42 " " + 4 " " " = 42 | = | 84·0 |
| Dividing by . . . . . . . . . . . . . . . | | 6)170·4 |
| | | 28·4 |
| Multiplying by the length of section . . . . . . . . . . . | | 100 |
| Solid contents . . . . . . . . . . . . . . . | | 2840·0 |

Secondly, if we suppose the upper surface to have the following form; viz., that of a plane passing through the points B C E, and intersecting in the line C E, a plane passing through the points C D E; the solid will then contain two apparent pyramids and two blind pyramids, viz., the apparent pyramids B *a* C E and D E *f* C, and the blind pyramids C *a f* D and E *f a* B; to measure which we have

| | |
|---|---|
| For the area of base of D E *f* C = 43·2 feet, and + 4 times middle area = 43·2 | = 86·4 feet |
| " " " *a* B C E = 42 " + 4 " " " = 42 | = 84·0 |
| " 4 times middle area of blind pyramid C *a f* D = C *a* × *f* D = 4 × 24 . . | = 96·0 |
| " " " " " " *a* B E *f* = *a* B × E *f* = 21 × 3·6 . . | = 75·6 |
| Dividing by . . . . . . . . . . . . . . . | 6)342·0 |
| | 57·0 |
| Multiplying by length of section . . . . . . . . . . . | 100 |
| | 5700·0 |

We thus find the upper portion of the second half section under consideration computed.

| | |
|---|---|
| For the 1st form of surface gives solid contents . . . . . . . . | = 2840 feet |
| " " 2d " " " " " . . . . . . . . | = 5700 |
| Difference in cubic feet . . . . . . . . . . . . | 27)2860 |
| " " " yards . . . . . . . . . . . . . | = 106 |

The wide difference in the results shows the necessity of noting while in the field the form of the twisted surface, that the proper method of computation may be applied. But it is not supposed that surfaces like those we have been considering will very frequently occur in practice, as the engineer would be likely to divide the section into two or three, which would have a tendency to lessen the differences much. But, to repeat, if the form of a surface be noticed when the field work is performed, so that the proper form of dividing the cross section may be applied, large sections may frequently be computed with as good a degree of accuracy, or even better, than smaller sections without such notice.

To complete the computation of the cubic contents of the figure, we have

| | | |
|---|---|---|
| The double area of the base of pyramid H M *f* D *a* . . . . . . . . . | = | 330 feet |
| Also " " " " " *a* B L I H . . . . . . . . | = | 240 |
| Then, 4 times the middle area of the blind pyramid = *f* M × B *a* = 21 × 10 . . | = | 210 |
| And 4 " " " " " " " = B L × M H = 9 × 8 . . | = | 72 |
| Dividing by . . . . . . . . . . . . . . . | | 6)852 |
| | | 142 |
| Multiplying by . . . . . . . . . . . . . . . | | 100 |
| Cubic contents of *a* B L I H M *f* D . . . . . . . . . . . | = | 14200 |

Having thus discussed the operations necessary to obtain the cubic contents of every portion of the section, we now add an example of summing up the contents after the manner in common practice. Before entering on our work, we remark that we contemplate two summations; the first containing the computation of the upper portion of the second half section, noticed in the foregoing as containing no blind pyramid; and the second containing the computation of the upper portion of said second half section, noticed as containing two blind pyramids.

**(66)** Now, as we have the areas computed, and marked upon the diagram of the cross sections, as described in the foregoing, we have as follows:

| | | First Summation. | | Second Summation |
|---|---|---|---|---|
| | No. 1 30·00 feet taken twice = | 60·00 feet | | 60·00 feet |
| | " 2 48·00 " " " = | 96·00 | | 96·00 |
| | " 3 58·8 " " " = | 117·60 | | 117·60 |
| | " 4 120·00 " " " = | 240·00 | | 240·00 |
| Blind pyramid | " 1 E *g* × A *c* = 15 × 5·6 = | 84·00 | | 84·00 |
| | " 2 *g* M × A *c* = 15 × 8 = | 120·00 | | 120·00 |
| | " 3 L *c* × G M = 4 × 9 = | 36·00 | | 36·00 |
| | " 5 42·00 feet taken twice = | 84·00 | | 84·00 |
| | " 6 120·00 " " " = | 240·00 | | 240·00 |
| | " 7 43·2 " " " = | 86·4 | | 86·4 |
| | " 8 165·00 " " " = | 330·00 | | 330·00 |
| Blind pyramid | " 1 = M *f* × B *a* = 10 × 21 = | 210·00 | | 210·00 |
| | " 2 = B L × M H = 8 × 9 = | 72·00 | Blind pyramid | 72·00 |
| Dividing by | . . . . . . | 6)1776·00 | *a c* × *f* D = 4 × 24 = | 96·00 |
| | | 296·00 | E *f* × B A = 3·6 × 21 = | 75·6 |
| Multiplying by length of section | . . . . | 100 | Dividing by . . | 6)1847·6 |
| | | 27)29600·00 | | 324.6 |
| Cubic contents, according to 1st summation | = | 1096·30 yds. | Length of section = | 100 |
| | | 1202·22 | | 27)32460·00 |
| Difference | . . . . . . . = | 105·92 | Cubic yards . . | 1202·22 |

The figures considered in the foregoing pages are those most commonly met with in railroad excavations. We frequently meet with modifications, however, containing points necessary to be noticed, between the slope and centre stakes; but it is believed that the ingenious engineer will, with a little practical experience, be enabled so to arrange the division of the sections, whatever may be their form, into pyramids, so as to admit of a ready and satisfactory method of computing their solid contents.

Before we leave this subject, we would remark, that the general rule we have been endeavoring to demonstrate gives the measurement of the middle area of the blind pyramid C *a f* D in the second half section, a trifle large; but so near the truth, that it has not been deemed necessary to change or modify the form of division or computation. In the supposed figure, named above, which presents rather an uncommon case, the error is something less than a square yard. If the length of the section had been taken at some thirty, or even fifty feet, the error would have been much diminished.

We had thought of adding some formula for computing the cubic contents of what we technically term borrowing pits; but, as a great majority of the figures composing these pits possess forms or solids so simple in their character that they will, at first thought, suggest ready and appropriate methods of computation, we deem it unnecessary to further enlarge upon the subject.

www.ingramcontent.com/pod-product-compliance
Lightning Source LLC
LaVergne TN
LVHW011216110826
845150LV00006B/1446

* 9 7 8 1 4 2 5 5 1 6 8 1 9 *